Medical Innovation

This book brings together a collection of empirical case studies featuring a wide spectrum of medical innovation. While there is no unique pathway to successful medical innovation, recurring and distinctive features can be observed across different areas of clinical practice. Drawing on a broad body of research on Innovation, Science and Technology, this book examines why medical practice develops so unevenly across and within areas of disease, and how this relates to the underlying conditions of innovation across areas of practice.

The contributions contained in this volume adopt a dynamic perspective on medical innovation based on the notion that scientific understanding, technology and clinical practice co-evolve along the coordinated search for solutions to medical problems. The chapters follow an historical approach to emphasise that the advancement of medical know-how is a contested, nuanced process, and that it involves a variety of knowledge bases whose evolutionary paths are rooted in the contexts in which they emerge.

This book will be of interest to researchers and practitioners concerned with medical innovation, management studies and the economics of innovation.

Davide Consoli is Research Fellow at INGENIO (CSIC-UPV), a joint research centre of the Spanish National Research Council (CSIC) and the Polytechnic University of Valencia (UPV), Spain.

Andrea Mina is University Lecturer in Economics of Innovation, Judge Business School, University of Cambridge, UK.

Richard R. Nelson is Director of the Program on Science, Technology, and Global Development, Columbia Earth Institute, Professor Emeritus of International and Public Affairs, Columbia University, USA.

Ronnie Ramlogan is a Senior Lecturer at the Manchester Business School, UK.

Routledge international studies in health economics

Edited by Charles Normand
Trinity College Dublin, Ireland
and
Richard M. Scheffler
School of Public Health, University of California, Berkeley, USA

Medical Innovation

Science, technology and practice

**Edited by Davide Consoli,
Andrea Mina, Richard R. Nelson
and Ronnie Ramlogan**

Routledge
Taylor & Francis Group
LONDON AND NEW YORK

First published 2016 by Routledge

2 Park Square, Milton Park, Abingdon, Oxfordshire OX14 4RN
52 Vanderbilt Avenue, New York, NY 10017

Routledge is an imprint of the Taylor & Francis Group, an informa business

First issued in paperback 2020

British Library Cataloguing in Publication Data
A catalogue record for this book is available from the British Library

Library of Congress Cataloging in Publication Data
Medical innovation : science, technology and practice / edited by Davide
Consoli, Andrea Mina, Richard R. Nelson and Ronnie Ramlogan.
pages cm
Includes bibliographical references and index.
1. Medical innovations. 2. Medical innovations–History. I. Consoli, Davide.
II. Mina, Andrea. III. Nelson, Richard R. IV. Ramlogan, R.(Ronnie)
R149.M3628 2015
610–dc23
2015020121

ISBN: 978-1-138-86034-6 (hbk)
ISBN: 978-0-367-66868-6 (pbk)

Typeset in Times New Roman
by Deer Park Productions

Contents

    a treatment for heart failure**   48
    PIERA MORLACCHI AND RICHARD R. NELSON

    Introduction 48
    Contextual background 49
    *A detailed history of the origin and evolution of the
       Left Ventricular Assist Device (LVAD) 51*
    The evolution of the LVAD as a treatment for heart failure 65

4   **Uncertainty in the hybridisation of new medical devices:
    The artificial disc case**   69
    DAVID BARBERÁ-TOMÁS

    Introduction 69
    Uncertainty and the artificial disc 73
    Hybridisation and the artificial disc 76
    Conclusions 82

5   **Technological accretion in diagnostics: HPV testing
    and cytology in cervical cancer screening**   88
    STUART HOGARTH, MICHAEL HOPKINS AND DANIELE ROTOLO

    Introduction 88
    Method 89
    The development of cervical cancer screening in the USA 90
    The career of the HPV test 92
    Conclusions 110

6   **Poliomyelitis vaccine innovation**   117
    OHID YAQUB

    Introduction 117
    The emergence of a problem and the vision of a solution 118
    *Brodie-Kolmer vaccine failures: a weak testing regime in
       need of strengthening 120*
    The construction of a more sophisticated testability regime 121
    *Passive immunisation: testing for design and field-based
       capabilities 125*
    Killed vaccines: testing regimes for taking 'calculated risk' 127
    Live vaccines: testing in the shadow of the killed vaccine 131
    Discussion and conclusions 134

Figures

Tables

Contributors

David Barberá-Tomás Polytechnic University of Valencia, ES

Davide Consoli National Council for Scientific Research, ES

Stuart Hogarth King's College, UK

Michael Hopkins University of Sussex, UK

Andrew James University of Manchester, UK

Stan Metcalfe University of Manchester, UK

Andrea Mina University of Cambridge, UK

Piera Morlacchi University of Sussex, UK

Richard R. Nelson Columbia University, USA

Ronnie Ramlogan University of Manchester, UK

Daniele Rotolo University of Sussex, UK

Gindo Tampubolon University of Manchester, UK

Ohid Yaqub University of Sussex, UK

Preface

Stan Metcalfe

Who would disagree with the claim that the remarkable growth of standards of living in the Western world since the 1850s is closely connected to the growth of useful knowledge? Who would not recognise that these economies are instituted and organised as open systems, in which innovations of countless form repeatedly transform economic activity and the way lives are lived? That they are systems designed for the pursuit of creative destruction, as Schumpeter so aptly expressed the idea, is surely beyond doubt. Yet, our statistics of aggregate economic activity capture these phenomena to only a limited degree, not least because they hide the deep changes in economic structure that come with innovations and their spread. Moreover, they do not account for vital dimensions of the growth of human well-being, in particular, when considered in terms of the long-term decline in lifetime hours of work and the increase in life expectancy, surely one of the most remarkable conjunctions of modern economic progress. That we are expected to live longer and enjoy useful lives longer is very much due to a long sequence of innovations in medical practice and it is innovations in medical practice that are the focus of this volume of essays. They illuminate many modern themes.

It is fashionable to speak of the knowledge economy and to recognise that much of its activity is not about the production of goods in farm and factory but rather about the provision of services. If any sector has the claim to be a knowledge-based service activity it is medicine in all of its many forms.

Medicine, considered as a branch of the service economy, is concerned with the maintenance and repair of human kind and so, as a branch of engineering, it has much that is akin to the repair of any machinery. But humans are not standardised machines; the complexities of their functioning and the idiosyncratic features of each individual mean that the response of two patients presenting with the same symptom and experiencing the same treatment may be radically different. Part of success in medical innovation involves the extension of workable procedures to an ever wider range of patients, all notionally experiencing the same problems. Complexity and variations are two of the key aspects that challenge the progress of medical knowledge and practice as the essays in this volume so clearly establish.

As a branch of the knowledge economy there is more to be said about medicine. Modern medicine involves a deep division of labour between distinct specialities, each marked by their own practices and bodies of understanding. Exponents of each sub-specialisation may have little understanding in common when it comes to the knowledge that is decisive for their success in treating patients. The days of the generalist are long passed; indeed, many of the innovations reported on here have generated their own sub-discipline to serve as a locus for practice and the ongoing development of knowledge about specific ailments. From a wider perspective this is an economy of collective knowing but private ignorance and, like all systems premised on the division of labour, its functioning depends on how the components are organised and connected. We benefit from modern medicine because each practitioner is highly specialised: rather as in Adam Smith's famous pin factory, the division of labour is a record of problems solved and executed with greater efficacy by individuals who know a great deal about a very precise and narrow problem but are ignorant of the wider field.

Ignorance is, of course, the corollary of uncertainty, our 'unknowledge' as George Shackle once expressed it, and any innovation process is directed at the replacement of uncertainty by more secure yet fallible understanding. In this regard innovation is a class of problem-solving activity - the exploration and development of pathways to effective solutions. Many paths are tried but not all acquire the consensual feature that marks an effective solution to a particular problem. Because presently agreed solutions have the attribute of social technologies with their in-built instituting mechanisms, there may be open resistance to the promise of new pathways, not least because new understandings may render existing knowledge and practices obsolete. Moreover, some paths are driven through very rough and inhospitable territory, progress is slow and, as any mountain walker knows, the next step may serve only to reveal how unexpectedly far they are from a destination. The studies reported here make it abundantly clear that multiple kinds of knowing are involved in the development of each particular problem-solving sequence, and that these different kinds of understanding are complementary but they may often be deficient. There can be no simple formula to the effect, for example, that a breakthrough in basic science will solve the problem in hand; it may contribute and indicate new ways forward but, in practice, the solutions required are engineering-like solutions that work in practice on specific patients. Who leads and who follows in terms of kinds of knowing is a most unproductive way to think of advances in medical practice.

It is remarkable how all of the breakthroughs reported here were developed by what we might call a distributed innovation process, with multiple actors making different contributions premised on their own understandings. Such an innovation-systems perspective points to the importance of the ecology of contributing actors (firms, universities, hospital clinics), to the different incentives that motivate their behaviour, and to their connectivity and the many ways that connections are made and broken. As Schumpeter made clear, the prospect of profit is important where capital is to be invested, as it is in private firms, but other actors within the innovation system have quite different incentives. To say that one motive is superior to any other is a further unproductive way to approach the medical innovation process. It is remarkable how the mix of different kinds of knowledge, different organisations and different incentive systems is such a rich source of problem-solving capability. Indeed, although the contributors to this volume are economical in the policy implications they draw, the overarching implication is the need for openness in the organisation and instituting of medical practice. Our medical systems are susceptible to invasion by novelties so current practice may fall victim to the impossible rendered actual. Therefore, keeping the medical system open to novelty is perhaps the greatest task of policy. Perhaps there is a case for more museums of medical practice, just as there are museums of science and engineering – their purpose being not only to remind us that the past was different to the present, but also to ever alert us that the future will be different from the present. Perhaps this would serve to reinforce one of the principal messages of these fine essays, that progress involves an ever shifting balance of agreement and disagreement: medical knowledge and practice is a restless system, never entirely at ease with itself, always searching for the solutions to problems whether new or old. Long may it remain open.

Acknowledgements

Part of this work could have not been possible without the generous support of the following organizations: the UK Economic and Social Research Council, the UK Medical Research Council, the Marie Curie Fellowship of the European Commission, the Wellcome Trust, and the Ramon y Cajal Programme of the Ministerio de Economia of Spain. Too numerous to mention, we thank the countless individuals participating in the many conferences and workshops where we shared our work for their challenging comments and questions.

However those attending the following were particularly instrumental: 'The Uneven Growth of Knowledge' Workshop in Florence, organised by Richard Nelson and Fabio Pammolli in 2004; the 'Healthy Innovation' Workshop organised by the ESRC Centre for Research on Innovation and Competition, Manchester, UK in May 2004; the 'SPRU 40th anniversary' conference organized by SPRU, University of Sussex, Brighton, UK in September 2006; the workshop on 'Dynamics of Knowledge and Innovation in Knowledge-Intensive Industries' organized by BRICK/GREDEC at Collegio Carlo Alberto, Turin, Italy in December 2008; and special thanks must go to Professor Annetine Gelijns and colleagues for their kind invitation to participate at the Grand Rounds Presentations in the Department of Health Evidence and Policy at Mount Sinai School of Medicine in New York in October 2010. We are also grateful to François Perruchas for technical assistance in putting this book together.

On a final note, we wish to thank John Pickstone, a stern critic and supporter of our work. John, a world renowned historian of science, technology and medicine and founder of the Centre for the History of Science, Technology and Medicine based at the University of Manchester, passed away in 2014. In him we found a kindred spirit and an inspiration for our research.

Introduction

Sources and effects of medical innovation

Davide Consoli, Andrea Mina, Richard Nelson and Ronnie Ramlogan

Perspectives on innovation in medicine

This book brings together a collection of empirical case studies that highlight a wide spectrum of important, if inadequately recognised, features of advances in medicine, namely: the persistent uncertainty that permeates dealing with human disease; the importance of failure as a starting point for successful innovations; the role of the institutional context in opening up or hampering new ideas and methods; the persistence of inefficient practices; the leadership of key individuals and the supporting role of emergent professional networks; the reconfiguration of scientific boundaries; and the role of conflicts within and across professions. Taken together, the chapters that make up this volume show that, while there is no unique pathway to successful medical innovation, recurring and distinctive features can be observed across different areas of clinical practice. Common to all the case studies is that the efforts to innovate were a response to perceived weaknesses in prevailing patient care. The cases also highlight that not all problems have solutions within the prevailing state of knowledge, and that workable solutions often prove to be of very differing efficacy. The key question is therefore why medical practice develops so unevenly across and within areas of disease and over time, and how this relates to the underlying conditions of innovation across areas of practice.

The focus of the studies contained in this volume is on the evolution of diagnosis and treatment within specific disease areas. Some of the innovations analysed here fit within existing modes of practice, and thus might be regarded as incremental, while others have more radical effects on the organisation of health care delivery and may even lead to further sub-specialisation. All the empirical material presented in the ensuing chapters directs attention away from isolated inventions and focuses instead on the pathways that underpin the transformation of medical problem sequences. One of the purposes of this volume is to bring together a series of contributions that disentangle the nexus between the technological, institutional and organisational dimensions of health innovation systems.

Why a book on medical innovation?

The subject matter of this book, medical innovation, is of interest for a number of reasons and to a wide readership. To begin with, it clearly is important to understand how many diseases and scourges that were the prime causes of infirmity or death before, say, World War I have been eliminated or got under strong control in large portions of the industrialised world (McKeown 1976). Some of these remarkable advances owe much to general improvements in living standards such as quality of housing, nutrition and sanitation, as well as enlarged access to healthcare. However, in a large number of circumstances reduced mortality and incidence are due to enhanced medical practices. As Nelson *et al.* (2011) forcefully argue, this progress cannot be ascribed to a single cause but is rather the result of various processes. This picture is further enriched by the stark contrast between continual success in some areas of medicine and the stubborn lack of progress in others. Questions about why certain diseases have not been mastered, at least not yet, deserve attention, and part of the explanation lies with gaining a better grasp of how innovation in medicine works.

An additional element of relevance of the present volume is the distinctive character of the subject matter. Medical innovation is a discovery process that makes impossibilities possible through the application of skill (*techne*) and knowledge (*episteme*). What is different about innovation in medicine, compared with innovation in other fields of activity, is that the design space of medicine is the human body, a complex structure for which parametrisation is not easy in the event of a malfunctioning. Designing solutions to human disease entails confronting two forms of uncertainty: first, illness is a physiological process that is hard to isolate in laboratory conditions; and, secondly, restoring health is commonly pursued with the aid of chemical compounds or mechanical artifacts on the basis of the expectation, generated by past experience, that a solution will trigger the desired reaction in the organism. But if there can be reasonably good understanding about 'what kind' of reaction can be expected, much less exists on 'how much', 'for how long', and the full range of effects that may follow. Put otherwise, health problems are generally detected 'on-line', that is, in the course of clinical practice, but the search for a solution is often carried out in a controlled, 'off-line', environment. A central question therefore concerns how reliable information originating in such an environment can be for actual practice, where a lot of non-controllable factors exert a decisive influence (Gelijns 1991; Nelson 2003). Add to the above that the penumbra of uncertainty looming over medical innovation, both understood as 'technique' and as 'body of scientific understanding', extends even further when ethical dilemmas enter the equation. The history of medicine is ripe with advances stemming from experimentation carried out in conditions and circumstances that by today's standards would seem unacceptable or forbidden. No doubt, controversial cases persist to this day.

Captivated by the kaleidoscopic challenges associated with providing health in a way that is both effective and viable, social scientists have devised various interpretative and analytical models to elucidate different aspects of medical practice. However, little agreement has emerged on the nature of innovation in this complex realm (Consoli and Mina 2009). Our specific approach to this topic entails a broad definition of medical innovation that encompasses combinations of new drugs, devices and clinical practices deployed in the provision of healthcare (Djellal and Gallouj 2005; Pammolli *et al*. 2005). Contrary to narrow definitions focused solely on artifacts, this book takes a broader view of the goods and services as well as of the social technologies and institutional structures that support the emergence and diffusion of improved ways to restore health. The organisational and institutional arrangements that underpin the development of new medical technologies over time are central to the view proposed here.

Building on the foregoing, the present book aims to be a point of reference for an emergent community of scholars. Partly due to the influence of earlier publications of the studies contained in this volume, the subject matter has attracted growing attention in recent years. So much so that medical innovation enjoys a steady presence at conferences, often at dedicated special sessions, in scientific journals and is also the central theme of several doctoral and postdoctoral research projects. This outbreak of interest takes the cue, both conceptually and methodologically, from an established scholarly tradition of specific industry studies aimed at elucidating the determinants of technological, organisational or institutional innovation (see, for example, Mowery and Rosenberg 1999; Mowery and Nelson 1999; Murmann 2003, to name a few). Following on from this, we will now explore the commonalities and the differences between innovation in medicine and in other fields that have been studied.

Characteristics of innovation in medicine

The roots of scholarly research on medical innovation lie arguably in the policy discourse of the 1950s when technology creation and diffusion were often conceived of as a linear process stemming from basic research and unfolding through to adoption and use (Bush 1945).[1] As a result of the ample body of empirical work that has been produced since that time, the intellectual support for this model has progressively waned. For one thing, the depiction of the route between R&D and technology adoption as a linear model neglects the influence of end-users who have been observed to be important for articulating needs and devising alternative ways to cater for them (Von Hippel 1976). Furthermore, the direction and the timing of invention engender significant unevenness across areas of expertise (Rosenberg 1976; Nelson 2003) and this poses formidable obstacles to both the diffusion of medical technology (Serra-Sastre and McGuire 2009) and to the translation of medical knowledge into practice that the linear model does not account for (Cooksey 2006). Last but not least that approach overlooks the negotiated nature of many technologies, prominently including medicine, and the uncertainty involved in the production and legitimisation of new know-how within and between professional groups (Rosenberg 1989; Pickstone 1993; Webster 2002).

The discontent with the established tenet inspired a new strand of scholarly work in the 1990s (Blume 1992; Gelijns and Dawkins 1994; and Rosenberg *et al*. 1995). Gelijns and Rosenberg (1994) first put forth the notion that diagnosis and treatment in medicine advanced through incremental adjustments to counter limitations that emerge only in actual practice. The model of medical knowledge that took shape in this new perspective exhibits clear commonalities with the way in which scholars of technology portray innovation in the context of engineering (see, for example, Constant 1980; Vincenti 1990), in particular the circumstance that medical innovations stem from trial-and-error experimenting rather than being guided by sharp scientific knowledge of disease. As a matter of fact, due to the dispersed nature of knowledge and the imperfect nature of communication, much of the relevant know-how is generated at the interface of neighbouring institutional domains such as the clinic, the industry and academia (Consoli and Mina 2009).

Conventional wisdom regarding the sources of progress in medical practice has emphasised the role of scientific research in deepening knowledge of how the human body works and of the nature of disease. Indeed few would disagree that new understandings won through research paved the way to many improvements in the ability to diagnose and cure but, at the same time, few appreciate that scientific research is just one of the forces that fosters progress in medicine, while other important pathways have gone largely unnoticed (Nelson *et al.* 2011). One is associated with advances in technology not originally conceived with medicine as the main context of use, such as electronics and computing as well as advances in new materials. Their appearance was independent of medical knowledge but ended up playing a pivotal role in the development of many new drugs and devices thanks to the proactive role of skilled practitioners. Learning in clinical practice is another important source of progress in medicine, for at least two reasons. First, whether a technique or artifact works as predicted can be judged only in circumstances that can reveal its actual strengths and weaknesses. Second, effective use of new devices often requires the development of dedicated procedures and organisational routines. Thus there tends to be significant interaction between what is learned in the context of use and the adjustment of the artefact. In a number of cases this iteration between learning in practice and learning in research has proven crucial for innovation.

Besides enriching the scholarly debate, this new wave of research has also revived the health policy discourse. For example, Christensen *et al.* (2007) warn about the dangers of sweeping generalisations in the evaluation of medical devices and drugs. The reputed gold standard, Randomised Clinical Trials, is a useful source of evidence on the safety, the efficacy and the effectiveness of new medical technologies, but it is also a setup that does not fully account for the peculiar challenges that characterise clinical practice and, *a fortiori*, for the learning that may originate from it. Studies on innovation in medicine indicate that repeated observation of what happens in actual use, and not just the results of formal testing, is needed to assess reliably the efficacy of a practice (Ligthelm *et al.* 2007). On the other hand, the firm grounding of innovation studies in appreciation of historical processes fuels a critical stance towards the deterministic design of large-scale policy programmes. Echoing the view of several scientists, most agree that solutions to complex problems are unlikely to come about 'by directive' (Mowery *et al.* 2010). The discoveries of vaccination, of insulin and of the aspirin are powerful instances of how effective solutions stem from adaptive problem-solving. In the words of Groopman (2000), had a centrally directed programme tackled the problem of polio, it is likely that we would have developed 'highly sophisticated iron lungs instead of the vaccine that eradicated disease'. All in all, errors and unintended consequences are crucial, if unpredictable, ingredients of progress in medicine, and there are great benefits from a thorough understanding of how and why technology fails; of why some scientific conjectures persist in guiding clinical practice in the face of evidence that they are not (or only partially) correct; of why some inventive efforts succeed more quickly than others; and, finally, of why successful breakthroughs fail to generate the same benefits across different health systems. To be sure, the policy logic that stems from this view does not advocate the design of systems that prevent errors altogether because innovation processes are uncertain and inherently wasteful. Rather, it is advocated that a health system should allow enough room for experimentation, and that it desirable to have in place pathways that facilitate the detection and the diffusion of successful emergent practices.

The empirical studies that make up this volume draw on and contribute to the conceptual framework just outlined. Innovation is portrayed as a trajectory of improvement sequences that extend the range of application of procedures, improve practice, and that often challenge existing know-how and change it. Underpinning it all is the accumulation of knowledge that, as new unforeseen hurdles emerge, calls upon different types of practitioners who carry experiences and competences and fuel different visions. The distributed nature of medical innovation entails that problems are solved and created by multiple actors within different organisations guided by distinct incentive systems (Kline and Rosenberg 1986; Nelson 2003). This perspective highlights the importance in innovation studies of recognising both 'medical innovation ecologies', namely the set of individuals working within repositories and generators of new knowledge, and the 'system making' connections between the components that ensure the flow of feedback. Quite crucially, any innovation ecology is the basis for a system but it is not a system itself until the relevant actors connect intentionally to combine multiple sources of knowledge (Consoli *et al.* 2009). Health innovation systems typically are cross-sectorial, international and characterised by mixed governance structures (Consoli and Mina 2009).

Who are the relevant actors of these systems? Clearly the division of labour between complementary public and private domains in healthcare is an intricate matter, and the influence of the former on the innovation ability of the latter varies depending on the industry (see, for example, Cohen *et al.* 2002), the disease area (HERG 2008) or the country (see, for example, ESF 2007). Therefore, at the forefront are healthcare providers such as hospitals, loci of clinical practice and the major channel through which new treatments reveal their latent potential as well as their drawbacks. These actors hold the key to the formulation of new ideas stemming from both existing techniques and serendipitous observations (Djellal and Gallouj 2005). In recent years the contribution of research hospitals has been praised as a valuable source of user feedback (Von Hippel 1988; Roberts 1989). Owing to their expert knowledge in articulating medical needs, these mixed entities have become key hubs for linking experimental phases of research (i.e. the clinical trial) and basic science (Gelijns and Thier 2002). Business firms constitute another prominent pole in the constellation of agents involved in the search for novel solutions in medicine. Producers of pharmaceutical products or of artifacts account for large shares of overall investments in high-technology products across most advanced economies. Over the last few decades the list of technology areas in which these firms engage has expanded significantly to encompass electro-medical equipment, surgical and medical instruments, surgical appliances and supplies, as well as drug design and production. Medical technology companies carry distinctive, and often global, capabilities in product development, and strive to facilitate the translation of advances in specialised areas, such as microelectronics, telecommunications and biotechnology, into the clinical realm. Given the uncertain nature of devising and implementing medical solutions, these firms often opt for a model of value creation that focuses not merely on the efficient production of devices or drugs, but also on the smooth integration of their new products into the continuum of healthcare. This task requires the participation of other actors such as, for example, healthcare deliverers (as per the above) or the institutions in charge of the regulatory process for the approval of new drugs and devices. Furthermore, medical technology firms rarely venture out by themselves but, rather, brave the uncertainty of exploration by setting up collaborations with universities and specialised organisations in charge of 'basic' research that is broadly oriented and is thus potentially beneficial for various fields of application (Nelson 1982; Griliches 1992). Research organisations, the third big type of actor in the medical innovation universe, are often public institutions that undertake a variety of roles including dissemination activities, supply of basic and specialised training, consolidation of clinical guidelines, promotion of health lifestyles, and the development of prototypes. Contrary to an established perception, over the last few decades the public sector has played a more than active role in supporting 'applied' research and development, especially in the United States (Sampat 2011).

The literature above suggests clearly that, for all the merits of individual actors in achieving breakthroughs, progress in medicine is the output of an evolving ecosystem. All the studies contained in this volume strive to capture and analyse the interface across different institutional boundaries, not as a datum, but rather as a context-dependent lever for unleashing scientific and technological learning.

A brief characterisation of the cases

This collection of case studies adds to the existing literature on medical innovation in different ways. Each chapter offers a historical perspective on the development of a particular clinical area and builds from it a critical overview of the sequences of problems and solutions that define the path of medical practice in the context at hand. These studies therefore differ from other works on medical innovation in that they do not merely recount the 'career' of any specific artifact (see Blume 1992) but rather seek to capture the changing boundaries of the surrounding medical discipline as it goes through successive clinical and scientific modalities (Pickstone 1993). In so doing, the chapters appreciate the intertwining of several interconnected domains, namely: scientific understanding; institutional frameworks; technological infrastructure; organisational set-ups; and the influence of changing cultural perceptions of disease and of ways to tackle it.

The first case study describes the development of the Intra-Ocular Lens (IOL), a clinical breakthrough that has radically transformed the design and the delivery of a clinical treatment of cataracts through the replacement of the impaired eye lens with an artificial functioning lens. The achievements of both the artifact, the lens, and of the surgical procedure for its insertion in the eye are the result of progressive systematisation of best practices, pioneered by creative individuals, which became the basis for the establishment of a transnational medical-industrial complex and an associated service delivery system. The second case reflects on the development of a technical device for the treatment of Coronary Artery Disease (CAD), namely Percutaneous Coronary Intervention (PCI) or coronary angioplasty. Percutaneous Coronary Intervention is the story of unforeseeable challenges in the clinical realm giving way to problem-solving sequences that spanned complementary realms of knowledge, technology and institutions. The third case study recounts the origins, the development and diffusion of the Left Ventricular Assist Device (LVAD), a pumping device that is implanted in the body to restore functionality in patients with insufficient cardiac output. This chapter illustrates the co-evolution of multiple forms of know-how by means of feedback mechanisms across clinical practice, medical research and engineering design. The stories recounted in the first three studies share a distinctive feature, namely that online experience with the main devices enabled better understanding of the disease. The fourth case study on the artificial intervertebral disc for Degenerative Disc Disease (DDD) tells a different story and illustrates the peculiar characteristics of competing artifact designs, each responding to a specific logic, that is, to differing interpretations of how the disc is supposed to work and, accordingly, of how the artifact is expected to alleviate pain. In the face of persistent uncertainty on what works and what does not, the emergent clinical solution has been the convergence of competing designs into a dominant hybrid artifact.

The common element across these initial chapters is that the evolution of artifacts has spurred progressive adaptations to the delivery of the attendant clinical services. In some cases this has elucidated new aspects of the aetiology of disease, while in others it has not. The remaining contributions deal with somewhat different aspects of medical innovation. The fifth case study focuses on the institutional factors that shaped the appearance of a new diagnostic modality for the screening of cervical cancer in the US, the human papilloma virus test. The co-existence of this innovation with the traditional pap smear test exemplifies the changing logic of diagnostic innovation driven by a reconfiguration of professional boundaries as well as of the competences and incentives carried by increasingly prominent private actors. The sixth case study explores the story of vaccine innovation to combat the poliomyelitis virus and illustrates the intertwining of technical problems and institutional adaptations driven by trial-and-error. This overview highlights the concomitance of elements such as testing through intermediate conditions, development of skills and instruments, and adaptive governance models. The final case study deviates somewhat from the path marked by the preceding ones in that it covers a clinical territory significantly plagued by lack of understanding and by the persistent uncertainty of the effectiveness of known solutions. 'Glaucoma' is a group of ocular disorders with unknown aetiology leading to permanent loss of sight if untreated. Some forms of this disease are hard to detect or too aggressive for effective prevention and, even in the case of timely diagnosis, the existing therapeutic regimes can slow down but not reverse visual field loss. Against this backdrop the chapter addresses the question: how does the problem-solving strategy of a community of practitioners unfold in the face of persistent uncertainty? This restates the overarching theme of the present volume, namely that what can be learned off-line about the causes of disease is often of limited help for actual therapy due to the uncertainty that characterises the pursuit of progress in medicine.

Note

1 See Edgerton (2004) and Balconi *et al.* (2010) for discordant views on the linear model.

References

Balconi, M., Brusoni, S. and Orsenigo, L. (2010) 'In defence of the linear model: an essay'. *Research Policy*, 39(1), pp. 1–13

Blume, S. (1992) *Insight and Industry: On the Dynamics of Technological Change in Medicine*. Cambridge, MA: MIT Press

Bush, V. (1945) 'Science: the endless frontier'. *Transactions of the Kansas Academy of Science*, 48(3), pp. 231–264

Christensen, M. C., Moskowitz, A., Talati, A., Nelson, R. R., Rosenberg, N. and Gelijns, A. C. (2007) 'On the role of randomized clinical trials in medicine'. *Economics of Innovation and New Technology*, 16(5), pp. 357–370

Cohen, W., Nelson, R. R. and Walsh, J. (2002) 'Links and impacts: the influence of public research on industrial R and D'. *Management Science*, 48: pp. 1–23

Consoli, D. and Mina, A. (2009) 'An evolutionary perspective on health innovation systems'. *Journal of Evolutionary Economics*, 19(2), pp. 297–319

Consoli, D., McMeekin, A., Metcalfe, J. S., Mina, A. and Ramlogan, R. (2009) 'The process of health-care innovation: problem sequences, systems and symbiosis' in J. Costa-Font, C. Courbage and A. McGuire (Eds) (2010) *The Economics of New Health Technologies: Incentives, Organization, and Financing*. Oxford: Oxford University Press

Constant, E. W. II (1980) *The Origins of the Turbojet Revolution*. Baltimore, MD: Johns Hopkins University Press

Cooksey, D. (2006) *A review of UK health research funding*. Report, available online at: http://www.hm-treasury.gov.uk/media/4/A/pbr06_cooksey_final_report_636.pdf

Djellal, F. and Gallouj, F. (2005) 'Mapping innovation dynamics in hospitals'. *Research Policy*, 34(6), pp. 817–835

Edgerton, D. E. H. (2004) 'The linear model did not exist' in K. Grandin, N. Worms and S. Widmalm (Eds.) *The Science-Industry Nexus: History, Policy Implications.* Sagamore Beach, MA: Science History Publications, pp. 31–57

European Science Foundation (ESF) (2007) *European Medical Research Councils: Present Status and Future Strategy for Medical Research in Europe.* ESF Strasbourg

Gelijns, A. (1991) *Innovation in Clinical Practice: The Dynamics of Medical Technology Development.* Washington, DC: National Academy Press

Gelijns, A. C. and Dawkins, H. V. (Eds) (1994) *Adopting New Medical Technology.* Washington, DC: National Academies Press

Gelijns, A. C. and Rosenberg, N. (1994) 'The dynamics of technological change in medicine'. *Health Affairs* 13(3), pp. 28–46

Gelijns, A. C. and Thier, S. O. (2002) 'Medical innovation and institutional interdependence'. *JAMA: the Journal of the American Medical Association*, 287(1), pp. 72–77

Griliches, Z. (1992) *The Search for R&D Spillovers* (No. w3768). National Bureau of Economic Research

Groopman, J. (2000) *Second Opinions: Stories of Intuition and Choice in the Changing World of Medicine.* New York: Viking Penguin Group

Health Economics Research Group (HERG), Office of Health Economics, RAND Europe (2008) *Medical Research: What's it Worth? Estimating the Economic Benefits from Medical Research in the UK.* London: UK Evaluation Forum

Kline, S. J. and Rosenberg, N. (1986) 'An overview of innovation' in R. Landau and N. Rosenberg (Eds) (1986) *The Positive Sum Strategy: Harnessing Technology for Economic Growth.* Washington, DC: National Academies Press

Ligthelm, R. J., Borzì, V., Gumprecht, J., Kawamori, R., Wenying, Y. and Valensi, P. (2007) 'Importance of observational studies in clinical practice'. *Clinical Therapeutics*, 29(6), pp. 1284–1292

McKeown, T. (1976) *The Modern Rise of Population.* London: Edward Arnold

Mowery, D. C. and Nelson, R. R. (Eds) (1999) *Sources of Industrial Leadership: Studies of Seven Industries.* Cambridge, UK: Cambridge University Press

Mowery, D. C. and Rosenberg, N. (1999) *Paths of Innovation: Technological Change in 20th-century America.* Cambridge, UK: Cambridge University Press

Mowery, D. C., Nelson, R. R. and Martin, B. R. (2010) 'Technology policy and global warming: why new policy models are needed (or why putting new wine in old bottles won't work)'. *Research Policy*, 39(8), pp. 1011–1023

Murmann, J. P. (2003) *Knowledge and Competitive Advantage: The Coevolution of Firms, Technology, and National Institutions.* Cambridge, UK: Cambridge University Press

Nelson, R. R. (1982) 'The role of knowledge in R&D efficiency'. *The Quarterly Journal of Economics*, 97(3), pp. 453–470

Nelson, R. R. (2003) 'On the uneven evolution of human know-how'. *Research Policy* 32, pp. 909–922

Nelson, R. R., Buterbaugh, K., Perl, M. and Gelijns, A. (2011) 'How medical know-how progresses'. *Research Policy* 40(10), pp. 1339–1344

Pammolli, F., Riccaboni, M., Oglialoro, C., Magazzini, L., Salerno, N. and Baio, G. (2005) *Medical Devices Competitiveness and Impact on Public Health Expenditure.* Bruxelles: Entreprise Directorate-General, European Commission

Pickstone, J. V. (1993) 'Ways of knowing: towards a historical sociology of science, technology and medicine'. *The British Journal for the History of Science*, 26(04), pp. 433–458

Roberts, E. B. (1989) 'The personality and motivations of technological entrepreneurs'. *Journal of Engineering and Technology Management* 6, pp. 5–23

Rosenberg, C. E. (1989) 'Disease in history: frames and framers'. The Milbank Quarterly, 67(1), pp. 1–15

Rosenberg, N. (1976) *Perspectives on Technology*. Cambridge, UK: Cambridge University Press

Rosenberg, N., Gelijns, A. C. and Dawkins, H. (Eds) (1995) *Sources of Medical Technology: Universities and Industry*. Washington, DC: National Academies Press

Sampat, B. (2011) 'The impact of publicly funded biomedical and health research: a review' in *Measuring the Impacts of Federal Investments in Research: A Workshop Summary*. Committee on Measuring Economic and Other Returns on Federal Research Investments, The National Academies (Appendix D)

Serra-Sastre, V. and McGuire, A. (2009) 'Diffusion of health technologies: evidence from the pharmaceutical sector' in J. Costa-Font, C. Courbage and A. McGuire (Eds) *The Economics of New Health Technologies: Incentives, Organization, and Financing*. Oxford, UK: Oxford University Press

Toole, A. (2007) 'Does public scientific research complement private investment in research and development in the pharmaceutical industry?' *Journal of Law and Economics* 50, pp. 81–104

Vincenti, W. G. (1990) *What Engineers Know and How They Know It: Analytical Studies from Aeronautical Engineering*. Baltimore, MD: John Hopkins University Press

Von Hippel, E. (1976) 'The dominant role of users in the scientific instrument innovation process'. *Research Policy* 5(3), pp. 212–39

Von Hippel, E. (1988) *The Sources of Innovation*. Oxford, UK: Oxford University Press

Webster, A. (2002) 'Innovative health technologies and the social: redefining health, medicine and the body'. *Current Sociology* 50(3): pp. 443–457

1 The intra-ocular lens revolution

Stan Metcalfe, Andrea Mina and Andrew James

Introduction

Worldwide, it is estimated that some 39 million people are blind while over 240 million suffer from low vision. Around 33 per cent of these cases are due to the presence of cataract. The cost of this form of visual impairment is immense in terms of human functioning and wellbeing, and in terms of lost economic output in advanced and developing economies. In terms made familiar by Amartya Sen, the activity of the restoration of sight brings the promise for great improvements in human capabilities, improvements not captured accurately by conventional measures of economic output (Sen 1999). In this chapter we trace the development of an extremely important medical innovation in the field of ophthalmology, the intra-ocular lens, a radical innovation that transformed the treatment of cataract (Apple *et al.* 2000).

We interrogated the technical literature, collected and analysed data on the emergence and diffusion of the technology, and interviewed expert informants (including clinicians, surgeons and hospital managers).[1] The method is comparative and historical and combines qualitative and quantitative data extracted from a variety of primary and secondary sources. Beside the relevant medical literature and the interview materials, we consulted national surveys on technology diffusion for the US and UK. We then constructed two datasets, one of papers and one of patents, by keyword searches and used them to complement the analysis of the epistemic evolution of the problem sequence as profiled in appreciative accounts. The paper dataset includes around 2,300 papers extracted from the Institute of scientific Information (ISI) covering the period 1965–2000. The patent dataset contains 707 documents extracted from the US Patent Office and covers the period 1976–2002. Finally, we used institutional sources (OECD, FDA and NHS documents[2]) to investigate the regulation of ophthalmologic practice, the creation of demand and the nature and constraints of the adoption decision.

The innovation of the Intra-ocular Lens (IOL) has radically transformed the conception, design and delivery of a major medical service; the removal of cataracts combined with their replacement by a functioning lens. This has brought great benefit to countless patients and has greatly increased the efficiency and effectiveness with which the clinical procedure is carried out.[3] It has been achieved by the creativity of individual clinician inventors combined with the development of a transnational medical-industrial complex that has changed radically the innovation system in this field of ophthalmic medicine. A procedure originally based around pioneering 'hero-surgeons' deploying 'craft technique', has evolved into a 'routine, quasi-factory' procedure capable of being effected in a local medical centre by clinician nursing staff, whose education and training have correspondingly changed, and who are supplied with high quality lenses from competing companies.[4] This is indeed a fundamental transformation of a service activity, its skill base and its innovation dynamic.

This chapter is structured as follows. In the second section we briefly outline the cataract condition, the 'problem sequence' that the IOL 'solves' and the innovative vision of the pioneer of the technique, Harold Ridley. In the third section we trace the evolution of the problem sequence from the work of Ridley's followers to the revolution of foldable lenses. In the fourth section we highlight how the emergence of correlated understanding shapes the dynamics of the micro innovation system. In the fifth section we explore the relevance of demand and regulation in the development of IOL. The sixth section concludes the chapter.

Innovation and the problem sequence

Cataracts and their treatment

Cataracts, the clouding of the eye's crystalline lens (the terms dates back to the Middle Ages and the intended imagery is of the descending of turbulent water in the eye), are the most frequent cause of defective vision in later life. The restriction of the passage of light is progressive and results in blurred vision, colour distortion and glare disability in the presence of bright light. Ultimately resulting in blindness, cataracts are severely disabling for otherwise active people. Between the ages of 52 and 64 there is a 50 per cent chance of their occurrence but by the age of 75 some 70 per cent of the population suffer from cataracts. With an ageing population in the world, the significance of an effective cure is not easily overestimated.[5] Surgical treatment of cataracts dates back to the Middle Ages, and possibly to Egyptian times, and until the late 1940s the possible operative techniques remained unchanged in their essentials. Couching, the process of physically displacing the cataract within the eye, was the only available 'treatment' until Daviel developed the first operative, extractive technique known as extracapsular extraction (ECCE) in 1748. This method removed the cataract nucleus but left the capsular bag *in situ* in the eye. An alternative technique, known as intracapsular cataract extraction (ICCE), dates from the late nineteenth century, and removed the lens, together with the lens capsule in which it is located. The relative advantage of these two operative techniques plays an important role in the intra-ocular lens story. These techniques were handicraft methods, and used various designs of knife to make incisions in the eye and extract the cataract.

Success depended entirely on the skill of the surgeon and the care with which the implements were sterilised. A successful operation meant that light could now pass to the retina but without the patient having the lens with which the light is focused as a clear image (the condition known as aphakia). The only corrective method was to use external lenses, in the form of 'pebble glasses', thick and unwieldy lenses, which give poor post-operative vision, magnifying and distorting images (Kaufman 1980). The risk of infection and of collateral operative damage meant that this was a procedure of the last resort for most patients. Moreover, the procedures involved extensive hospitalisation and months of post-operative recovery. Periods of up to 160 days are reported for a leading centre in the late nineteenth century (Schlote *et al.* 1997), and as late as the 1940s a hospitalisation of three weeks or more could be anticipated. Only the effectively blind could expect to benefit from this style of treatment. Thus, removal of the cataract was at a price and to describe this operation as an ordeal in which defective vision is replaced by defective vision seems entirely appropriate (Linebarger *et al.* 1999). This is the state of ophthalmology in relation to cataracts when Harold Ridley emerges as an inventive force.

Harold Ridley and the intra-ocular lens

At the time of his invention, Ridley was the senior eye surgeon at Moorfields Hospital in London, the then leading eye hospital in the UK. Appointed a full eye surgeon and Fellow of the Royal College of Surgeons, at the early age of 32 (1938), Ridley soon established himself as a medical educationalist of distinction, introducing many reforms into the curricula for the training of eye surgeons. After war service in Africa and the Far East, he returned to Moorfields and established himself as an inventor in many areas of ophthalmology.[6] However, it is in relation to IOLs that Ridley transformed the treatment of cataracts. In this respect he is a classic inventor/innovator in the Schumpeterian mould, the relatively isolated individual generating and applying a new combination of ideas in the face of substantial hostility.[7]

Ridley's contribution was four-fold: it entailed the creation of a new surgical technique, the design of a new implant, the use of new materials and the development of new equipment. We might say, following Usher (1929), that he followed a process of cumulative synthesis to realise his invention. As we pointed out above, in the first half of the twentieth century the dominant operative method followed by cataract surgeons was the ICCE procedure in which the entire lens is removed within its capsular bag.

Ridley had formed the view that ICCE was a technique inferior to the alternative of extra capsular cataract extraction (ECCE), in which the lens capsule is left *in situ*, and his insight involved inserting the plastic lens into the capsular bag. Crucially, he believed that the prevailing lens extraction procedure was 'but halfway to a cure which is complete only when the lost portion is replaced' (Ridley 1951, p. 617).[8] The choice of PMMA (acrylic) as the lens material is an instructive example of the role of the unexpected in the innovation process.

Wartime injuries to pilots had indicated that Perspex 'shrapnel' would lie 'inert' in the eye, producing minimal pathological or chemical reaction.[9] Here then was an ideal material, inert and light (almost the same specific density as eye fluid). Yet, industrial Perspex clearly would not do as it contained too many impurities so, lacking the requisite chemical knowledge, Ridley joined forces with John Pike of Rayner (an ophthalmological supply company) and John Holt of ICI Ltd to develop 'Perspex CQ' (Clinical Quality), a suitably purified form of PMMA. Ridley, Pike and Holt, worked in secret on the use of ECCE technique, the design of the rigid lens, its manufacture from PMMA (Perspex), and the insertion of the lens in the posterior chamber of the eye. The three worked together on a non-commercial basis (for fear of the wrath of fellow clinicians) and Rayner agreed to manufacture the lens and supply them on a cost-only basis. Here we find the first tentative shaping of a local distributed innovation process, bringing together the complementary capabilities of the clinician, the technician and the industrial chemist.

Ridley implanted the first IOL in a 45 year-old woman on 29 November 1949. The operation took place at St Thomas' Hospital London, using the ECCE technique (the removal of the cataract predated the insertion of the implant by three months). His second implantation took place in August 1950. Ever aware of the radical nature of his multi-dimensional invention, Ridley delayed until July 1951 before he announced the results of his operations to the annual ophthalmic conference at Oxford, the premier meeting of the British ophthalmic community. By then, 17 months after his first procedure, he had implanted eight Rayner-manufactured lenses. Not unexpectedly, the response from leading eye surgeons present at the Oxford meeting was almost uniformly hostile.

Technical hurdles and opposition

The nature of the professional hostility to this innovation is an intriguing aspect of the story and it is not entirely irrational. Ridley's IOL was a double innovation in terms of conception and surgical procedure, it was a radical alternative to established practise and it challenged an established viewpoint that cataract extraction using ICCE was the best that could be achieved. As a new technique, it placed great demands on the skill of the surgeon and created major risks during and after the operation.

In the early years, the first lenses were too thick and heavy.[10] Furthermore, they were turned by hand and, consequently, varied from copy to copy. No method existed for sterilising the lens, thus post-operative inflammation posed severe limitations to the success of the new surgical practice.[11] Dislocation (the slippage of the implant out of the line of sight), primarily due to damage to the posterior capsule or zonule during surgery, constituted another major problem that was not solved satisfactorily for many years. As a result of these complications, some patients subsequently had to have their lenses removed.[12]

Cataract surgery is not a theoretically grounded science, theory does not predict how an individual patient will respond to any method, so it is not entirely unreasonable that experience should dominate the world-view of its practitioners or that professional reaction is conservative. Another factor is important here. A medical innovator will know that his technique will involve risks for the immediate patient that may only become clear over an extended horizon. In putting the development of the technique above the interests of the immediate patient, the innovator is invoking an abstract notion of patient welfare in general. When the technique is established others will benefit but the moral dilemma is clear - the cost of progress for many may be born directly by individual patients. This is precisely the dilemma that the rules and norms of the profession are meant to deal with and these rules, as accumulated social capital sunk in the profession, will constrain and channel the acceptance of new methods to make life difficult for innovators. Equally, no account can properly ignore the many patients willing to endure the risks of a new and experimental treatment and so benefit future generations.

Thus we find the eminent Duke-Elder, in the 1959 edition of his authoritative 11-volume text, noting that the novelty and difficulty of the technique had led to few surgeons using it (Duke-Elder 1959, Vol. 11, p. 289) and he concluded with this assessment:

> it is perhaps unwise to gamble on further surgical procedures which require considerable specialised skill and a healthy eye on which to operate the results of which in the absence (at present) of longer-term observations are somewhat problematical. (p. 291.)

Although Ridley had his followers, among which at least Peter Choyce in the UK and Edward Epstein[13] in South Africa are worth mentioning, there is no reason to doubt that this negative assessment was generally held within the ophthalmological community of the time. Indeed, by then, Ridley was close to abandoning the implantation of IOLs.[14]

What is significant is that Ridley's innovation did ultimately sweep the world and Ridley was awarded one of the last Knighthoods of the 2nd Millennium for 'pioneering services to cataract surgery'.[15] By the end of the twentieth century the IOL had become the standard complement to cataract surgery, which itself had become one of the most frequently performed outpatient operations in the advanced industrial world (Linebarger *et al.* 1999). A major survey of the histopathology of IOLs opines that, 'lens implantation is among the safest major procedures in modern surgery' (Apple *et al.* 1984).

Yet in the mid-1960s it appeared to be a dead end and it is doubtful if Ridley or anyone else could foresee the steps that would transform the invention into a widely adopted innovation. So, how and why did this transformation take place? How did the implantation of IOLs progress from a local method practiced in multiple ways by different surgeons to a virtually uniform, universally applied technique? Part of the answer is provided by the emergence of a community of IOL practitioners. With little support from the industry, usually in terms of the limited manufacture of their idiosyncratic lens designs, this group of enthusiast 'hero-surgeons' formed the basis of a series of highly localised micro-innovation systems, introducing new variants on a trial and error basis and communicating the outcomes in the professional literature, at conferences and in visits to their respective medical centres.

The eye as a 'design space' and the emergence of a practitioners' community

IOLs are not unique in the fact that Ridley's radical invention and innovation has spawned a sequence of incremental innovations and the development of a range of complementary technologies. IOLs are a textbook case of the latent potential implicit in a radical innovation and the identification of a design space that could be explored by other innovators. Indeed, it is only with this subsequent process of exploration that the radical nature of the innovation became manifest.

One of the key factors in lens' design is its intended location in the eye and the different surgeons were, in effect, contesting the ideal design location. In many respects, the history of this innovation is a history of the eye being explored as an engineering design space. Figure 1.1 illustrates the main options, which are three in number: anterior chamber lenses, angle supported in the corner of the anterior chamber; posterior chamber lenses located in the capsular bag; and iris-supported lenses fixed in front of or behind the pupil.

The original Ridley lens was located in the capsular bag within the posterior chamber of the eye and relied on the ECCE operative technique. This was a risky procedure that led to a number of complications, including displacement of the lens, post-operative opacification of the posterior capsule and iris atrophy from contact with the optic. These complications, together with the demanding nature

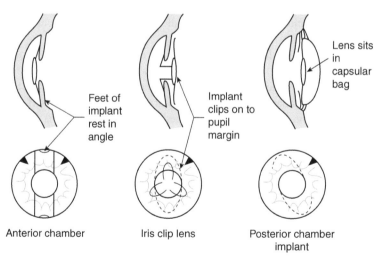

Figure 1.1 The eye as a design space: lens implant configurations.

Source: Wybar 1984.

of the technique, encouraged surgeons to experiment with lenses placed in the anterior chamber, the first of which was implanted in 1952 by Baron, and was followed by many other designs.[16]

Rival lens designs to those promoted by Ridley and his followers were developed by other leading eye surgeons. The most significant development here was the placement of the lens in the anterior eye chamber, a different portion of the design space. Strampelli, working in Rome, implanted the first widely accepted anterior chamber lens in 1953. Other clinicians followed his lead, including the eminent Barcelona-based surgeon, Barraquer, although, by the time he presented his results at the Oxford Conferences of 1956 and 1959, it was clear that the anterior chamber lens was creating new design problems. The size and curvature of the lens was crucial to its success, yet no accurate method existed for measuring the magnitude of the anterior chamber. Consequently, problems arose from the rigid lens touching and irritating the inner surface of the cornea, the endothelium. The resulting corneal dystrophic effect was to undermine the case for anterior chamber lenses but, for a while, they were the ascendant design (Barraquer 1956; 1959).[17]

The other major development in design was the introduction of iris-supported lenses; the first, in 1953, was Epstein's 'collar-stud' lens, followed by Binkhorst's 'iris-clip' lens in 1958. The logic behind the iris-clip design was the desire to avoid a major complication of posterior chamber lenses - their propensity for dislocation, and of anterior chamber lenses - the damage they inflicted on the cornea. As Binkhorst expressed the point, 'Therefore I designed an implant which, in a harmless way, is entirely supported by the iris diaphragm and does not touch the angle at all, nor other related structures' (Binkhorst 1959, pp. 573–574).[18] Between June 1956 and the Oxford presentation of 1959 he had carried out 19 implants and he clearly considered that his design was a major advance on the then prevailing alternatives. However, despite the initial promise and success of this design, long-term, multiple complications in relation to the stability of the lens and iris reaction have led to their eventual abandonment.[19]

It is clear from the above account that the years immediately following the announcement of Ridley's invention and innovation had stimulated a great deal of creative endeavour to contest and explore the design space of the eye.[20] Indeed, the Deputy Master of the 1959 Oxford Ophthalmological Congress was moved to suggest, when commenting on Binkhorst's paper, that 'it would appear that the subject of the IOL was taking the place of the former interest in glaucoma'.

The ferment of inventive activities that were emerging around lens insertion techniques in the early 1960s led to the foundation of the International Intra-Ocular Club. Meeting first in London in 1966, it formed the basis for the identity of the early community and subsequently became the European Society of Cataract and Refractive Surgeons, which now contains over 2,500 members in 100 countries and publishes its eponymous journal. Within this community, many other 'hero-surgeons' emerged as inventors cum innovators seeking to improve on Ridley's design.

The role of formal and informal interactions taking place within the framework of this community scarcely needs be emphasised. As is rather systematically the case for practice-based disciplines, such as surgery, communication of tacit knowledge via personal connection and direct interaction is pivotal for the resolution of ignorance and for medical progress to occur.[21] Ophthalmology is no exception. It is largely by personal, tacit and often unrecorded cooperation that individual knowledge comes to be more highly correlated and is built into a body of shared understanding. Shared understanding contributes to the formation of standards and influences the institutional framework in which trial and error experimentation is translated into accepted norms of practice. However, to take this innovation from within its 'hero-surgeon' community required much more than the activities of a professional society; it required further distributed innovation of technique, innovation that created a step change in the possibilities of application. It is to the sequence of innovations complementary to Ridley's that we now turn.

The evolution of the problem sequence

The next steps

No innovation takes place or diffuses in isolation and the determinants of success for medical procedures often reside in the development of complementary procedures, drugs and devices. This is certainly the case of the IOL. Of all the developments that have transformed Ridley's innovation and operative method into a mass procedure, by far the most important has been the adoption of phakoemulsification techniques for cataract extraction. This radical and complementary process innovation has enabled a step change in the operative procedures. It has stimulated the development of new kinds of foldable lenses to capitalise on the fact that a much smaller incision is made in the eye, with the effect that smaller wounds heal more quickly. Within the unfolding problem sequence, it triggered advances that placed the evolution of the IOL on a new, more vigorous, path; phakoemulsification is the most significant of process inventions. Most substantially of all, phakoemulsification has led to the routinisation of cataract extraction and raised the prospect of cataracts being removed by trained nursing staff. The bottleneck represented by the delicate skill of the surgeon was to a degree replaced by standardised, mechanised and replicable practice, and the economics of the procedure transformed as the operation became possible as an ambulatory procedure performed in a few hours.

As with Ridley's breakthrough, the diffusion of phakoemulsification (phako for short) has depended on a sequence of supporting innovations, including the use of viscoelastic agents and the introduction of new anaesthetic methods. Phako also ensured the dominance of the posterior chamber location in the eye and the return of the ECCE method, so favoured by Ridley, which had fallen under a cloud in the intervening years. Improved surgical microscopes and aspiration systems have also reinforced the position of phako methods despite the development of cryogenic extraction methods in the 1960s. The combination of developments was the key to safer surgery and improved postoperative life for patients, and they have turned cataract extraction into a new service industry in which cataract extraction is almost invariably followed by the insertion of an IOL.

The trigger invention: phakoemulsification

The originator of this technology was Charles Kelman, a Professor of Clinical Ophthalmology in the USA (New York). Kelman had established his credentials as an inventor in the 1960s, with the development of a sophisticated cryoprobe for the removal of cataracts. In 1963, he turned his attention to the question of the benefit to patients of a procedure that would reduce the size of the incision in the eye. His attempts to develop rotating mechanical cutting devices bore no fruit until, by chance, he realised upon a possible solution in an ultrasound device.[22] He experimented for many years with the idea of using ultrasound, that is to say the high frequency energy of a vibrating needle to fragment a cataract, which would then be sucked clear of the eye through a much smaller incision (2–3 mm) than that traditionally associated with the ECCE technique (10–11 mm).[23] Experimenting with the technique first on animals and cadavers, he carried out the first human operation in his private practice in 1967 and, by 1972, he reported that he employed it on 84 per cent of his patients (Kelman 1973). Improvements followed quickly and the first crude machines were made available commercially in 1970, signalling the shift in the locus of leading edge of commercial cataract innovation to the USA.[24] The device was patented in collaboration with an engineer, Anton Banko, and consisted of the ultrasound needle, a supply of irrigating fluid, a pump to evacuate the debris from the liquefied cataract, and a control mechanism for the surgeon (Kelman 1991).[25]

In presenting his method to the British Ophthalmological Society in 1970, Kelman claimed that its main benefit was the dramatic reduction in time lost by the patient. Even with the best ECCE techniques of the day, 4–8 days would be spent in hospital with a 6-month recuperation period at home. With 'phako' the patient left hospital the day after surgery and could be fully active immediately and, he added, a contact lens could be fitted after 3–5 weeks. Indeed, it is not clear that Kelman did other than fit contact lenses to his patients. The professional literature soon carried papers by other surgeons who reported outcomes similar to those achieved with the ECCE technique (Hiles and Hurite 1973; Cleasby *et al*. 1974). Thus was born the technique that transformed cataract surgery.

The diffusion of the phako method

We have seen how Ridley's intention of replacing the ICCE technique with the ECCE alternative had failed, and that his preferred method of locating the IOL in the posterior chamber soon fell into disfavour. Yet, by the end of the twentieth century, ICCE is defunct in the advanced countries (although not in the developing world) and the posterior lens is the standard fitting. Much of the explanation for this change is found in the growth of knowledge about phakoemulsification and the matching pattern of its diffusion.

Tracing the spread of phakoemulsification is not easy, records are sparse and the fact that many procedures were and are carried out in private clinics makes it difficult to paint an overall picture. However, sufficient fragments can be assembled to give a broad assessment that suggests that the method took off, on the way to becoming the dominant operative technique in the USA, around 1983. As supporting evidence, we have survey and other data on the use of the new methods in the clinical context. Consider first the USA. As late as 1969, the predominant method of cataract extraction was the ICCE procedure, which usually required a 2–4 day stay in hospital. By 1984, the picture had changed considerably. A national survey (Dowling and Bahr 1985) reported that 64 per cent of surgeons used the ECCE method and that 75 per cent inserted IOLs in more than half their patients, while 66 per cent of surgeons inserted IOLs in 90 per cent or more of their patients.

Clearly, the growth of the IOL market was well established, roughly 20 years after Kelman's invention. Moreover, 30 per cent of the reporting surgeons performed ambulatory surgery, that is to say the patients were day patients, and we can infer that many, if not all, of these cases involved the use of phako methods. Nørregaard *et al.*'s (1998) more limited survey figures suggest that by 1991 the proportion of operations using phako in the USA had risen to 66 per cent. Similarly, Jaffe (1999) reported that phako was used in over 85 per cent of operations by 1994. More recent figures from the American Society of Cataract and Refractive Surgeons suggest that, in 2000, phako was the procedure of choice for 97 per cent of surgeons (Leaming 1998). The upshot is that, in the USA today, cataract surgery is the most frequently performed surgical operation for individuals over 65, with some 0.5 million procedures taking place in 1984 and 1.3 million procedures by 1996 (Apple and Sims 1996).

These changes in operative technique from the early 1980s onwards are also reflected in the shifting balance between preferred lens designs. However, the return of the posterior chamber lens and the simultaneous decline of iris-clip lens began well before the phako method became established.[26] The following Table 1.1 shows the rapidly changing proportions in which the different lens designs were used in the USA between 1978 and 1984 (Stark *et al.* 1984). The phako method also spread quickly to the UK, but only after a lag of some 10 years. By the early 1970s, Arnott reports that an estimated 7,000 phako operations had taken place worldwide and that he had carried out 40 in the UK (Arnott 1973). He later reported on the growth in the use of the technique at Charing Cross hospital in London. Of some 113 operations in 1973, only 21 used phako while, by 1976, this method accounted for 118 out of 138 operations (Arnott 1977). However, it seems that the practice did not spread beyond the few pioneers.

Apparently, the method of resource allocation in the UK health care system constrained the rate of adoption because of a fear among managers that the potential demand could not be met. Beginning around 1990 adoption did take off and it did so rapidly. A national survey in 1997, covering operations on circa 18,500

Table 1.1 Use of competing lens designs USA (%)

LENS TYPE	1978	1984
Anterior chamber	25	30
Posterior chamber	4	69
Iris fixation	52	<1
Iridocapsular	19	<1

Source: Stark *et al.* 1984.

patients, found that 77 per cent of operations in 100 UK hospitals used phako but with considerable variation in usage between hospitals, covering the range from 10 per cent to 99 per cent. The survey also found that 70 per cent were surgical day cases (Desai *et al.* 1999a).

Figure 1.2 shows the diffusion of phako in the UK in the 1990s. Here we see the decline of ICCE to negligible proportions by the early 1990s when ECCE 'ruled the roost'. Initially, phako displaced ICCE procedures but, from 1992 onwards, the substitution is against the handicraft ECCE method. By 1999, phako has risen to dominance in the UK health system. Data for Manchester Eye Hospital, a large teaching hospital, cast interesting local light on the national picture. In 1991, four phako procedures took place out of a total of 3,246 cataract operations. By 1996, the phako proportion had risen to 52 per cent and to 90 per cent by 1999 out of a total of 4,102 operations. Over the entire period since 1991 virtually all those cataract operations were followed by the insertion of an IOL.

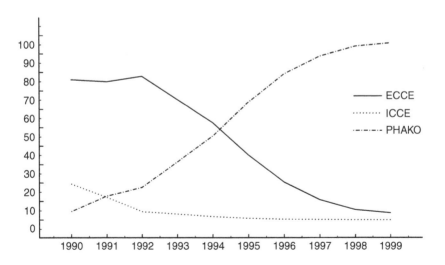

Figure 1.2 Diffusion of cataract extraction techniques in the UK.

Source: Manchester Eye Hospital records 2001, elaboration: authors.

Post-innovation improvements

No doubt part of the explanation of the slow initial diffusion of phako beyond a few adventurous surgeons lies in the need to acquire expensive suites of equipment combined with the need to acquire the necessary skills. Although Kelman provided short courses to aspiring phako surgeons, the immediate capacity to train and produce equipment must have placed supply side constraints on the rate of acceptance, independently of any natural conservatism of the ophthalmic profession. Nevertheless, there is more to the explanation than the slow adjustment of the forces of potential supply to latent demand.

Again, as with all radical innovations, the acceptance of this innovation depended on a sequence of post-innovation improvements to the technology. These were aimed at improvement of the control of energy levels, since the energy involved could do great damage if its application was not properly localised, and of the surgeon's ability to control the process. Improvements in the pumps and in the shape of the needle all played a part in improved performance and came through learning by using, as experience of procedures improved. However, these were engineering developments linked to a knowledge base in physics and electrical engineering. This is not the knowledge base of the typical clinician so the locus of invention and innovation necessarily moved to the firms competing in the market for these machines. Indeed what was happening in the field was the transition from the clinical invention system to a broader innovation system in which firms and clinicians play complementary roles.

The machine improvements aside, for phako to be accepted as the dominant operative technique a range of developments in other complementary technologies were also required. Two are of particular importance, and they illustrate the distributed nature of the innovation process and the emergent complementary relation between clinicians, scientists and engineers in the problem-solving process. In using the phako method, it is essential to stabilise the space of the eye and to hold constant the depth of the anterior chamber and the capsular bag in which the ultrasound energy is being dissipated. This is to prevent damage to the corneal endothelium and to the posterior capsule, either of which occurrence can cause serious post-operative complications. The widespread introduction of viscoelastics from the early 1980s solved these problems and their use has greatly improved the safety of cataract surgery techniques. The first use of viscous sodium hyaluronate was in 1972 and proved to be a turning point in the use of the phako method. Other viscoelastic substances soon followed and they are manufactured by all the major ophthalmic companies.

While this first example points to the significance of commercial firms in the innovation system, the second highlights the role of the creative surgeon. This second major advance is in clinical technique and is known as continuous-tear anterior capsulotomy. By this technique, the anterior capsule is torn by the surgeon in a circular fashion rather than being cut in a 'jagged' fashion as if with a can opener, hence its name. The advantage of this way of making the opening in the anterior lens capsule is that it prevents further rupture to that membrane when the operation is complete, and thus serves to stabilise more effectively than hitherto the location of the inserted lens in the eye. Two clinicians, Gimbel and Neuhann, introduced it independently in the late 1980s, and it has since become the standard method for opening up the anterior portion of the lens capsule.

Foldable lenses: a revolution complete

It is pointless to make a small incision with the phako technique if one has to make subsequently a larger incision to insert a conventional, rigid or semi-rigid PMMA lens. The development of the modern lens well illustrates the point that the solution of one problem often opens up a design space for new problems, so knowledge builds on knowledge in an autocatalytic fashion but, crucially in this case, the new knowledge lay beyond the ken of clinicians. The dominant solution in these problem sequences has seen the latest stage in IOL development, the innovation and adoption of foldable lenses. This innovation can be said to complete the revolution in cataract surgery begun by Ridley in 1949.

As with Ridley's lenses, the first generation of foldable IOLs were poorly manufactured and suffered many decentrations after insertion. Subsequent generations are thinner, have better haptics to stabilise the optic in the eye and have greater biocompatibility, as a result of greater understanding of the interaction between the new materials, acrylic and silicone, and the biochemistry of the eye. That firms are now the dominant players in the innovation system in this field is nowhere more apparent than in the design of the modern lens and the choice of material from which it is manufactured.[27] Much learning has occurred in the industry and, by the mid-1980s, Allergan, for example, introduced its three-piece silicone lenses, with UV filters incorporated in the material to prevent opalescence of the lens *in situ*. The second generation of silicone IOLs has further enhanced biocompatibility and they are thinner still. Side by side with the development of the new materials has been the development of new instruments to, for example, inject the foldable lens into the capsular bag.[28] Silicone is not the only new material made available. Alcon (a subsidiary of Nestle) markets acrylic foldable lenses (Acrysof). These lenses unfold more slowly within the eye, can be produced with a thinner optic and have many of the desirable attributes of PMMA.

Several patents have recently been issued for the injection of collagen-based compositions to form a new lens *in situ* and it is apparent that the materials' revolution is still in progress for IOLs. The most recent innovations have included the development of refractive, multifocal lenses that correct for short or long sight and, providing an alternative to radial keratotomy, the method of reshaping the eye through incisions or laser surgery. Yet further innovations are to be expected in materials and lens design in the future but here our account must stop, for the Ridley-inspired revolution is virtually complete, at least in the high-income countries.

The growth and transformation of medical knowledge in the IOL micro-innovation system

Composition, substitution and complementarities

While there is no obvious way to infer the development of knowledge in the minds of individuals, we can follow systematically the growing body of codified representations of personal knowledge placed in the public domain. This information, in the form of papers, patents, device evaluations and professional demonstrations of method, can provide invaluable insights into the development of correlated understanding within the community of practitioners and the supplying firms. It must be remembered, however, that placing information in the public domain is not placing knowledge in that domain, however tempting it may be to draw that equivalence. This is especially clear when knowledge, such as clinical/surgical knowledge, heavily depends on practice. This is in part the reason why we have, so far, emphasised practical cooperation, personal connections and informal exchanges. These appear to be especially important in phases of the early development of micro-innovation systems where knowledge is unstable and standards contested. At the same time, it cannot be denied that, as the system grows, the representation, communication and protection of private knowledge through publishing and patenting become essential factors in shaping the nature of further knowledge, as well as the scope for its future applications.

In the following section of the work, we integrate the evidence gathered in our interviews and in the relevant medical literature by examining how the growth and transformation of clinical and technological knowledge for the treatment of cataract is reflected in the codified traces of research activities contained in medical papers and patent documents. We examined around 2,300 scientific papers listed by ISI between 1965 and 2000 and 707 patents granted by the US Patent Office between 1976 and 2002. By means of simple statistics, we profile the main trends and explore the nature and change of composition, substitution and complementarity effects among the creative efforts of medical scholars and inventors. The main goal of this exercise is to provide evidence of the complex epistemic nature of the problem sequence. The following results corroborate the qualitative findings so far presented.

Figure 1.3 shows the proportions of all papers in the field of IOLs and cataract surgery that refer to at least one of the three main surgical methods that can deployed for cataract extraction procedures: ICCE, ECCE and phako. To begin, note that the trends broadly reflect actual practice in hospitals and clinics as reported in Figure 1.2. This is not insignificant because it constitutes clear evidence of the interdependency and co-evolution of medical research and clinical trial-and-error practice. The steady rise of papers referring to phako methods from the early 1970s is immediately apparent, as is the relative unimportance of references to ICCE, which was the well-established and thus uncontroversial practice of the profession. References to ECCE, Ridley's preferred method, begin to increase in the late 1970s, peak around the late 1980s, and then decay away as phako continues

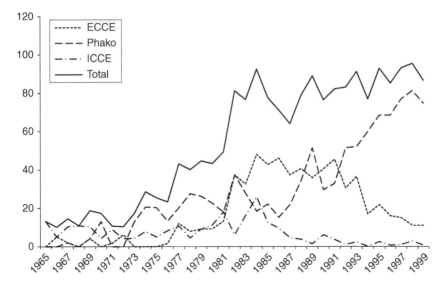

Figure 1.3 Distribution of IOL research publications 1965–1999.

Source: ISI, elaboration: authors.

its rise in importance. The joint growth of ECCE and phako-related papers reflect the shift to placing the lens in the posterior chamber, 'where nature intended'.

The growth of understanding clearly accelerated in the late 1970s, exactly in phase with the beginnings of rapid diffusion of the IOL into clinical practice. If we add together the three categories of papers charted in Figure 1.3, we have the proportion of the total of IOL and cataract surgery papers that refers to operative technique. In the late 1960s this total was less than 20 per cent of all papers, the inference being that most papers were then concerned with matters such as lens design or reports of complications once lenses have been fitted. The interest in matters of technique then rises almost year on year and accounts for 90 per cent of published papers by 1999. To put this information in perspective we have calculated the total number of papers in cataract surgery over the period since 1947. At that date they run at an annual rate of circa 150 papers per annum, rise to a peak of 350 in 1960 and settle down to a more or less constant rate of 250 papers in the early 1980s.[29]

The interest in matters of technique then rises almost year on year and accounts for 90 per cent of published papers by 1999. As the field of IOL-based surgery has grown, so the number of prolific authors of research papers has increased markedly. Take the relatively tough standard of a minimum of ten papers per annum produced over a ten-year period. In the decade, 1960–69, no author reached this level; the Russian ophthalmologist, Fyodorov, coming closest with six papers, followed by Binkhorst with five. By 1980–89, the picture has changed considerably. There are 26 researchers that meet this standard and, by the next decade, the figure has risen to over 40. At the same time the

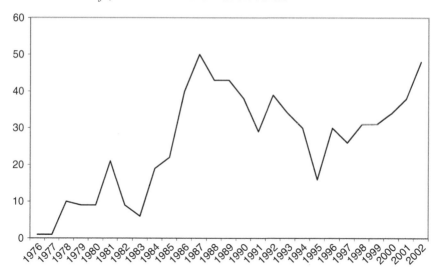

Figure 1.4 IOL patents 1976–2002.

Source: USPTO, elaboration: authors.

population of prolific authors working from the USA has declined, reflecting the international spread of the community of practitioners. By the 1990s, less than half the most prolific authors are operating from the USA compared with 80 per cent in the 1970s.

Patents offer a second window on the development of understanding with the added dimension that they reflect the growth of ideas with potential commercial value. Many of the pioneering IOL surgeons, as we have already suggested, patented their lens designs and operating instruments but, over time, there have been significant changes in the role of the surgeon inventors; as the community of practitioners has grown, a sub-division of specialised IOL surgeons has emerged and an increasing proportion of firms account for the majority of the patents granted. Figure 1.4 shows the trend in US patents for IOLs from 1976, following the date of invention of Kelman's invention, to 2001. We can see the considerable acceleration that took place after 1983, roughly corresponding to the emergence of commercial innovation systems.[30]

One of the interesting shifts that occurred in the community over this period appears to be the weakening association between the publishing and patenting activities of the top clinicians. In the early stages, prolific clinicians were also prolific patentees, Fyoderov and Kelman being prime exemplars, and the following Table 1.2 show the papers and patents associated with the ten most prolific publishers of papers from 1960 onwards.

The division of labour between 'research' and 'invention' has clearly changed; the old link between clinical practice and experimentation, publishing and patenting has been displaced as the innovation system has developed. The prolific

Table 1.2 Patents and papers of the ten most prolific publishers

	1960–69	*1970–79*	*1980–89*	*1990–99*
Papers	42	100	185	303
Patents	101	69	5	12

Source: Web of Science.

publishers in the new community of practitioners do not patent like the pioneers and this reflects the much greater role of commercial companies in the invention process. This shift towards commercial inventive effort is also reflected in the shifting balance of patent activity. Both changes appear to occur in the early 1980s, more precisely *circa* 1983.

The patent record also allows an assessment of the shifting balance of inventive effort as a further check on the evolving problem sequence. Beside its aggregate growth, we have identified the composition of patenting activity in the field of IOLs. Table 1.3 shows the distribution of patents across four categories, lens design, materials used in the making of lenses, methods of performing cataract surgery and tools, primarily for inserting foldable lenses into the eye. The proportionate growth in patents on methods and tools reflects the trigger effect of phako, and the decline in the relative importance of lens' design patents suggests that this dimension of the problem sequence is relatively settled.

Figures 1.5 and 1.6 show the extent of substitution effects in the composition of lens characteristics.

They illustrate respectively the rising relative importance of patents in relation to posterior chamber lenses, confirming the switch to the 'Ridley model', and the proportionate increase in patents concerned with deformable lenses.[31] It may be interesting to notice that, while the problem of lens location seems settled in 1995–1997, it in fact re-emerges from 1998 (although the number of location-related patents decreases in absolute terms). Examination of the patents granted after 1998 reveals that research on alternative locations was triggered by the need

Table 1.3 Distribution of patents by category

	No.			
	DESIGN	*MATERIALS*	*TOOLS*	*METHODS*
1976–1983	41	5	12	8
1984–1995	223	53	68	59
1996–2002	67	24	86	61
	%			
	DESIGN	*MATERIALS*	*TOOLS*	*METHODS*
1976–1983	62	8	18	12
1984–1995	55	13	17	15
1996–2002	28	10	36	26

Source: USPTO, elaboration: authors.

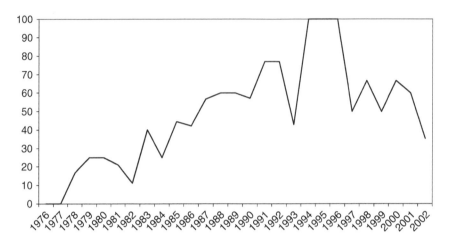

Figure 1.5 Proportion of posterior chamber lens 1976–2002.

Source: USPTO, elaboration: authors.

to accommodate the impossibility of placing lenses in the preferred location of posterior chambers because of local tissue damage. In this long-term development process, there clearly are connected and co-evolving aspects of the problem sequence where advances in one component are strongly associated with advances in the others.

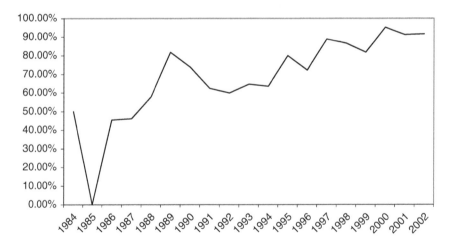

Figure 1.6 Proportion of patents classed as deformable 1984–2002.

Source: USPTO, elaboration: authors.

Division of knowledge, division of labour and interdependencies

Thus, Ridley's revolution was a revolution within the design space of the eye with multiple strands of invention and innovation required for its realisation. Like many transformative innovations it has been a long time in the making. The correlated growth of patents shows the very considerable shifts over time that took place in the body of correlated knowledge across this community. The move to foldable lenses placed in the posterior chamber by means of the phako method is one interrelated aspect of the same revolution in which the solution of one problem opens up new problems that find their solution in the inventive division of labour. The clinicians and firms in the supporting industry know different things; otherwise they would all invent in the same way, yet their knowledge is sufficiently correlated for them to appreciate the significance of developments made by others and include them in collective practice. It is a division of labour, one that is strongly coordinated by the practices that define the field. This division of labour in the production of private knowledge and public understanding is paralleled by a second division of labour in the organisation of the delivery of the service.

Knowledge of a practical kind does not grow in a vacuum and so, in parallel with the shifting patterns of papers and patents, we find major changes in the way the clinical service is organised and delivered, in effect the emergence of a new division of clinical labour.[32] Indeed, it is the practice of intra-ocular surgery in its clinical context that is reported in the growth of papers and patents. There can be no doubt that the revolution in the technique of cataract surgery begun by Ridley, and as reinforced by Kelman, is as much a profound revolution in the delivery of a clinical service as it is a revolution in knowledge and understanding.

The consequence of the sequence of innovations has been a change in the organisation of clinical service at large. Patients have clearly benefited beyond measure. A procedure that at the beginning of the 1980s required a general anaesthetic and a 5-to-7-day period of hospitalisation using the ICCE method, could then take place with a local anaesthetic and a 1-to-3-day hospitalisation using Ridley's preferred ECCE method, an improvement of itself. Yet, with the adoption of phako, surgery is now carried out on a day basis under local anaesthetic and with a very short recuperation period. In some UK hospitals the time from admission to leaving the hospital after the operation has been reduced to some two to three hours.[33] As a result, both the number of beds and nurses absorbed by cataract surgery has declined very sharply.[34] The cost savings in terms of capital and labour are clearly considerable. When taken in conjunction with the decline in postoperative complications, following the return of posterior chamber lenses and the improvements in lens quality, the transformation in the service for the patient is precisely immeasurable. It is this transformation of service delivery that lies behind the great increase in cataract surgery in the past decade and the fact that it is made available to patients who have many years of active life in front of them.

Wider still are the changes in the relative roles of nurses and clinicians in the provision of service. In the UK, specialised ophthalmic nurses now carry out many of the pre- and post-operative procedures formerly carried out by clinicians. The semi-routine nature of the equipment-based phako technique now raises the prospect that nurse-consultants will displace surgeons altogether in the removal of cataracts and the insertion of IOLs. This is surely an unforeseen consequence of the path down which Ridley ventured, though it is a consequence that surely would not have surprised Adam Smith in 1776!

One of the most striking features of the evolution of the IOL is the growing interdependence between the deliverers of health-related services and the manufacturers of lenses and related ophthalmic equipment. This interdependence marks the transition from localised invention systems to the innovation system of today, where large firms dominate and channel the innovation process along commercial lines. This interdependency takes a number of forms: the role of ophthalmic surgeons in the innovation process; the role of ophthalmic surgeons in testing new designs for regulatory purposes; and, the role of pioneering ophthalmic surgeons in legitimising new designs and practices within the surgical community.

In the early days of the procedure it was the 'hero-surgeons', Ridley, Kelman, Fyodorov and many others, who drove the field forward with the fundamental changes in perspective embodied in the concept of the IOL. Trial-and-error clinical experimentation and growing clinical experience have been the forces behind both radical developments in IOL technologies as well as incremental developments within established design trajectories, and the medical clinic or hospital has been the primary locus for the design of lenses and the accumulation of experience from the earliest days of the industry. Even with the emergence of the medical-industrial complex, the manufacturers of IOLs and related equipment remain heavily dependent on clinicians as a source of inventions, although the balance of advantage is changing.

This highlights the fact that – in a professional community where peer opinion is critically important – the views of leading surgeons can have an important influence on the diffusion of practices and selection of designs. The companies recognise this fact and build it into their innovation strategies. The leading surgeons are courted. They are encouraged by companies to use their products and the views of the leading inventors are taken seriously. Companies sponsor events at the leading ophthalmology conferences and in those supplements to the leading journals in which the merits of the various products are debated. Good platform speakers are at a premium, particularly where they are held in high esteem by their peers and are regarded as influential within the professional community. Indeed, particularly in the US, it is an accepted fact that certain surgeons are retained by companies to both advise them on matters of design and development and to promote the particular company's products through reporting on their (favourable) experience with those products.

What is also worth emphasising is that, even during the period of what we call craft-based innovation, new developments depended upon these mutual interdependencies but on a much smaller scale. Thus, Ridley's pioneering innovation in surgical technique depended upon the ability of ICI to develop clinical quality materials and on Rayner's ability to manufacture the lens itself – no trivial task in the late 1940s. Certainly, the importance of such informal learning and tacit knowledge in technological development in process equipment technology, particularly scientific instruments, machine tools and advanced manufacturing technologies, is well known.

The market for cataract surgery: demand, need and regulation

To the extent that the IOL is a story of increasing returns in the production and use of knowledge, we would expect that the scale of demand and the way demand is instituted play an important role in the unfolding of technique and practice. In this section, we explore the dynamics of the emerging IOL invention and innovation system and the forces in relation to the demand for IOL implants, the regulation of the practice and the development of commercial interests that shaped the system. This invention and innovation system did not exist naturally, it had to be instituted and the focus for its construction was the emerging of a series of problem sequences.

As with any dimension of economic life that is based on a division of labour, the growth in scale and composition of demand is a vital part of the story and the IOL story is no exception to this rule. The nature, size and growth of the market are fundamental determinants of the way healthcare is delivered and of the supporting innovation process. This is equally so in relation to cataract surgery and the implementation of innovations in IOLs. However, the market for treatment has its own peculiarities, expressed in the way that the participants make decisions and the rules, formal and informal, that govern the activities involved. In this case, four factors interact to determine the demand for the IOL procedure. These are the population at risk, the clinical procedures and routines that translate physical need into economic demand, the wider regulatory rules for the procedure, and the instituted norms for allocating health care resources to IOL treatment.

It is clear that the determinants of the pool of need and the rate of surgical removal of cataracts will vary from country to country but that, in all cases, the principal driving factor in adding patients to the pool is the age profile of the population. In the populations of the advanced countries, the incidence of cataract in the overall population lies in the region of 17–18 per cent and the predominant need for treatment is in those over 65 years of age.[35] That the population of advanced countries is growing older on average is an important factor in shaping the ongoing evolution of demand for cataract surgery. In the OECD region, for example, 13.3 per cent of the population in 1999 was over 65, a proportion up from 8.9 per cent in 1960. As per capita incomes have increased and the expectations of an active and long retirement have expanded so has the pressure from patients for this operation to be performed. At the same time, the more routine nature of the procedure and the increasing clinical confidence of positive outcomes means that patients in lower age brackets and with less acute conditions are operated upon.

While need is a physiological matter, its translation into demand depends also on the prevailing standard of visual acuity as assessed by professionals, and this in turn is related to their knowledge of and access to available techniques that work. The translation of patient need into demand for cataract surgery and the insertion of an IOL depend crucially on the prevailing clinical rules of assessment, and these in turn depend on the technologies of treatment open to clinicians. If the threshold of acceptable visual acuity is lowered this will expand, perhaps considerably, the pool of potential patients deemed to benefit from cataract surgery. A recent Australian study suggests that reducing the Snellen threshold from 6/12 to 6/7.5 would increase their pool of visually impaired candidates by a factor of 5 (McCarty *et al.* 1999). Similarly, in Denmark the number of annual cataract extractions increased by 350 per cent between 1980 and 1991, with a change in the surgical threshold cited as the principal source of this increase (Norregaard *et al.* 1996).

The advances charted above have transformed the need/demand relationship. One need only reflect on the hazards of the early procedures and the limitations of aphakic spectacles to see the point that, prior to the IOL, demand would necessarily be suppressed whatever the scale of need. Thus, one of the unforeseen developments, one that reflects the success of the technology, is the implantation of IOLs in patients who can expect many years of life with the lenses. Increasing life expectancy in the advanced countries means that an implant in a 50-year-old may be expected to function for another 20 to 30 years. Furthermore, intra-ocular lenses are today implanted in patients whose vision has been only marginally affected by cataracts, patients, for example, who are still able to drive a car.

A further factor affecting the growth of the IOL market is the particular ways in which countries fund medical services. In most of the advanced countries either medical care is funded by the state from general taxation supplemented with charges for medical prescriptions, as in the UK, or it is financed by charges on patients that are met from insurance payments, as in the USA. As has long been understood, these financial rules break the link between patient benefit and the opportunity cost of treatment and they have become something of a *cause celebre* in health economics. In state-funded systems, this typically leads to rationing of treatment with rationing decisions being in the hands of clinicians and medical system managers. In insurance-based systems, payments are typically set to cover the costs of treatment. In both cases, health professionals determine the relation between need and demand, and this is as true of cataract surgery and the IOL as it is of any other major area of treatment. However, the constraints that they face differ in the two cases.

In the UK, healthcare bureaucracies determined the timing of the expansion in demand, while in the USA it is the insurance companies that, through their rate-setting rules, incentivise physicians to perform particular procedures. In both cases pressure to cut costs has been an important factor in accommodating the growth of demand. As early as October 1984, the Medicare system in the state of Rhode Island was prescribing that all cataract/ IOL surgery should be carried out on an ambulatory basis to qualify for public reimbursement. No doubt, other States followed quickly and encouraged the adoption of phako techniques (Dowling and Bahr 1985). Similarly, in the UK in 1985, the NHS issued a directive to the effect that 80 per cent of cataract surgery should be performed on a day basis, thus encouraging the use of phako methods and the search for the efficient scheduling of operations.

Regulation of the market has been the third important factor extending the market for IOLs. Regulation is part of the instituted framework in which medical innovation takes place and, in the case of IOLs, the development of formal government regulatory frameworks paralleled but lagged behind the emergence of the medical industrial complex and this, in turn, lagged behind the self-regulation imposed by the clinical profession. In this sense the regulatory system, the market and the innovation sequence co-evolved. More forcefully, the development of regulatory institutions was essential to the growth of the market since it provided the assurance of a stable framework within which clinicians, patients, hospitals and suppliers could interact with a more developed sense of trust and correlated understanding. Effective and efficient devices and operative procedures provide benchmarks for quality and performance and facilitate the transfer of the understanding of practice to other surgeons. These standards do not exist naturally; they are discovered in an extended process of trial and error learning that reflects the engineering-like nature of the medical knowledge. Within this process, the creation of a sub-community of practitioners is a crucial step. Thus Ridley and his fellow pioneers had to create their standards through co-production with the activity.

The significance of this process is that concerns among the ophthalmology community about the safety and propriety of inserting a foreign body into the eye loomed large from the very beginning of the IOL story. Thus, we have noted that Harold Ridley and his hero-surgeon colleagues faced the criticism of Duke-Elder and the ophthalmology establishment in the UK, while the professional community – not least in the United States – was to long remain sceptical about the efficacy and ethics of the IOL. Indeed, this professional scepticism, backed-up by considerable clinical evidence of the problems and difficulties associated with the procedure, limited the diffusion of the IOL in the early years and provided the stimuli to improve the innovation. However, it was impossible to confine the norms and process of regulation within the ophthalmic community. Most significant of all was the emergence of a network of consumer activists in the USA, including Ralph Nader, which challenged the very basis of the ophthalmic community, namely its professional autonomy and the principle of self-regulation of practice and patient welfare.

Critics of the IOL procedure argued that IOLs had never been properly tested in animals or clinically investigated under properly controlled clinical trials. The consequence, they claimed, had been serious damage to the eyes of many patients including glaucoma, severe corneal disease, inflammation and infection. Trial, not surprisingly given the engineering nature of the medical knowledge, had resulted in error and the issue had become the regulation of practice to produce permissible error bounds. This challenge to professional codes of practice strained relationships within the ophthalmic community in the United States over an already controversial practice.[36] The direct consequence of this deep controversy was the extension of the regulatory powers of the FDA, in the 1976 Medical

Device Amendments to include IOLs, among other medical devices. The objective of these 1976 Amendments was to ensure the safety and effectiveness of medical devices by requiring manufacturers to register with the FDA and follow quality control procedures.[37] Paradoxically, it was the extension of the regulatory regime that was to underpin the growing acceptance of the IOL procedure in the medical community and by the medical insurance industry. Regulation no doubt constrained sharp practice but it also helped institute the market by adding an implicit minimum quality mark to the procedures, radically reducing uncertainty among patients and ophthalmic professionals alike.

This confluence of events, defined by an ageing population structure, a radically changed supply capability, and the regulation of practice and devices to make the market, underpinned the rapidly growing scale of practice and so provided the economic incentive for private firms to make large-scale investments in the cluster of techniques around cataract surgery and the insertion of IOLs. As reflected in the patent statistics, these changes came together in the early 1980s in the USA. In the process of making these investments, private firms have transformed a craft-based innovation system into a medical-industrial complex that transcends national borders. The crucial point is that the transformation of the relationship between an abstract concept of need and a concept of effective medical demand depended on the related transformation of a radical innovation into a routine medical procedure. Growth of the market stimulated the search for a new division of labour and this is reflected in the emergence of a modern micro-innovation system.

Conclusions

The IOL case is an important example of the interdependence between the service economy and the manufacturing economy. So close is the degree of supply chain interdependence that the medical service economy and the medical industry economy are effectively one, as the link between cataract surgery and the prophylactic use of IOLs serves to illustrate. Thus service innovation is premised on complementary innovations in manufacturing, and those manufacturing innovations are shaped by clinical innovations and related co-developments in the delivery of services. It is thus quite unhelpful to think of this economy as constituted by independent service and manufacturing sectors. What we are dealing with are knowledge-intensive medical services and the innovation systems that sustain them, and that transcend traditional sector boundaries linking the operation of trans-national medical device firms with the international community of clinical practitioners.

Linking the two levels are the external organisational arrangements of particular firms each with their own networks of suppliers and clinicians competing for business from hospitals and private clinics. Now no innovation system is formed without cost or without purpose. Rather, that division of labour is created and develops over time in order to stimulate the growth and application of knowledge and understanding. It is a distributed process of innovation that is continually evolving and shaping the accumulation of new knowledge and understanding, and much new knowledge accumulates in the context of 'market' processes.

There are two important, interrelated aspects of the process by which the IOL medical innovation system has emerged and developed. First, the level and attributes of national demand were important to the shaping of this dynamic division of labour. As we have seen, the relation between demand and need is itself dependent on the prevailing clinical technology and health care management practices. It is an instituted relationship. Secondly, the competitive activities of rival firms are central to the way the innovation system develops. The process of competition is reflected in the attempts of rival firms to build their own 'local' concentrations of innovation resources. That is to say, they develop proprietary micro-innovation systems as part of their strategies to support their ongoing search for competitive advantage.

Two consequences follow. We find competition to create different firm-specific systems of innovation, each one drawing its support from wider networks of knowledge-generating resources. Thus, competition between rival firms leads to contests for access to knowledge, expertise and skill held at national and sector levels of the ophthalmic medical innovation system. Secondly, to build firm specific innovation systems requires significant investments, not only in internal capabilities but also, for example, in the training of clinicians in the use of proprietary equipment and in the provision of clinical facilities. To build these relationships adds to the fixed costs of innovation and to the sources of increasing returns as the market for ophthalmic services expands. Hence, a key element in the development of the current configuration of leading firms is the large scale of the USA market for lens implantation.

The most important aspect of this case study is perhaps the fact that the innovation system has co-evolved with the development of technology and practice around specific problem sequences. It has been variable in form and geographic focus, variable in terms of the relative importance of surgeons and firms in the innovation process, and variable in relation to the understanding required to 'solve' the innovation problems of the moment. National institutions and organisations have played an important role in the sense of providing knowledge and other inputs into innovative activity and in framing the possibilities for innovation. However, national organisations did not form innovation systems because these depend on interaction for a purpose, and the linkages that define the pattern of interaction are constructed and friable. Thus, the development of IOL innovations is best understood in terms of evolving innovation systems focused around a constellation of problems related to one another and, at the same time, connected with the higher-order problem of the loss of human sight.

Acknowledgement

This chapter contains excerpts from Metcalfe, J. S., James, A. and Mina, A. (2005) 'Emergent innovation systems and the delivery of clinical services: the case of intra-ocular lenses' that have been reprinted from *Research Policy*, Vol 3, pp. 1283–1304, copyright 2005, with permission from Elsevier.

Notes

1 We thank Nick Jones, Michael Lavin, Karen Partington and Paul Habib Artes of The Royal Eye Hospital, Manchester, Heather Waterman of the School of Nursing, University of Manchester and Gary Young of The Manchester Central Health Trust for their very considerable assistance in the prosecution of this research.

2 Detailed references are in Section 4 of this chapter.

3 For the patient, an operation that formerly required months of incapacity is now recovered from in a matter of hours. For health services, there has been an enormous increase in capital and labour productivity associated with the increased patient throughput and the ambulatory nature of the modern procedures.

4 In effect, IOL implants have evolved into a commodity provided in a mass market, albeit a highly regulated one, that mixes public and private provision in different proportions according to country.

5 Although ageing is the major factor in cataract formation (senile cataracts), it is evident that their incidence is also influenced by lifestyle factors (smoking and alcohol consumption), the use of particular drugs (steroids), diabetes and exposure to UV radiation (West and Valmadrel 1995). Some cataracts in young people have genetic causes and occasionally the eye's lens has to be removed for other reasons as with traumatic injuries.

6 In 1948 he was the first surgeon to televise eye operations, and he devised an electronic system for examining the inner eye.

7 It is also worth emphasising that Ridley worked in the context of a healthcare system that provided key resources for his development activities, not least in the form of patients and the freedom to try his radical ideas without question. It is unlikely that a surgeon could proceed today with the freedom from bureaucratic scrutiny and imposed ethical constraint that Ridley enjoyed.

8 The plastic contact lens was invented in the 1930s as a device that was placed on the eye and was removable at will. Ridley's lens was also of plastic, though it was placed in the eye and was permanent. It is clear that Ridley was well aware of the design and use of plastic contact lenses and he was known to be turning them on a lathe in 1946 as an alternative to a moulding process (Ridley 1946). No doubt, his general inventive awareness provided him with important complementary knowledge.

9 Apparently, Ridley was mistaken in this conclusion in that most aircraft canopies were made of glass, which is equally inert. However, the general conclusion turned out to be correct.

10 Patel *et al.* (1999). A Ridley lens was 2.4 mm thick and weighed 108 mg, compared with the latest generation of lenses, at the time of writing, that are 1 mm thick and weigh 15 mg.

11 This problem was not solved until 1957 when Ridley introduced caustic soda as the medium for sterilisation.

12 Ridley considered that by 1970 this was necessary in more than 15 per cent of his cases. For any reasonable surgeon this outcome would be interpreted as an exceptionally high failure rate, and it would certainly colour the view of other surgeons against the technique. Ridley, however, held firm to the value of his methods, maintaining that they gave the patient a better visual field, an optically normal eye and avoided the problems of binocular rivalry.

13 E. Epstein was the surgeon who developed the Maltese Cross lens, also manufactured by Rayner. In the 1960s, he went on to develop soft implants, precursors of the hydrogel lens of the 1980s, and he was one of the early surgeons implanting foldable silicone lenses. On Choyce, see note 17.

14 In 1964, he assessed the history of his efforts in these negative terms: 'Risk of eventual failure with a posterior chamber lens is too great to be accepted though many highly successful results gratifying to both surgeon and patient have been achieved' (Ridley 1964, p. 7). Yet he concluded his assessment on a forward-looking, positive note, stating: 'In years to come this innovation in the surgery of the eye will find its true place' (*ibidem*, p. 13).

15 *The Times* of London for 31 December 1999.

16 Different lens designs were compatible with less-demanding techniques, and we find one surgeon suggesting that, by comparison, implantation in the anterior chamber 'was child's play compared to Ridley's technique' (Binkhorst 1959, p. 570).

17 Numerous developments occurred to find a solution to this problem, including lenses with more flexible haptics and lenses with open or closed nylon loops to lessen the irritation to the angle of the chamber (Dannheim and Barraquer designs). Choyce made important innovations in his search to improve the Strampelli lens, settling on a design with a rigid lens and flexible loops, and he worked closely with Rayner Ltd to develop the technology. In 1960, he reported that the improvements in the success of the IOL procedure were due to standardised machine-made lenses, better sterilisation methods, more experience in the design of the lens to fit the eye and in the choice of patient (Choyce 1960). Interestingly, two of his lens designs (the M8 and M9) were the first IOLs to gain approval from the FDA in the early 1970s.

18 The acrylic lens, only 0.6 mm thick, was placed in front of the pupil and held in place on the iris by wire loops. Also important as a designer of iris clip lenses was the Russian ophthalmologist Fyodorov, the pioneer of radial keratotomy. After correspondence with Epstein he also developed his own design of lens, the so-called 'Sputnik' lens.

19 Thus, it is reported that over the 4½ years to August 1982 the use of iris clip lenses in the USA declined from 52 per cent to 6 per cent of all lenses implanted (Apple *op. cit.*, p. 10).

20 It is important to notice that the competition between cataract surgeons over the best design and location of an IOL did not exhaust the logical possibilities for the treatment of aphakia following cataract surgery. The major competing alternative has already been alluded to above. During the 1950s and 1960s the use of and knowledge about the properties of PMMA contact lenses grew apace. Duke-Elder's text refers to their use, post cataract surgery, as easy and safe. Yet this apparently promising solution failed. Although many cataract extractions were followed by the prescription of contact lenses, for elderly patients (the bulk of the cases) they were considered unsatisfactory. A principal reason for this is alleged to be the difficulty that such patients experienced in using these lenses and in maintaining a satisfactory level of hygiene. In other cases, for example, the loss of a lens in a younger person due to trauma, a more convincing case could be made for contact lenses. Contact lenses are an interesting part of the IOL story for it is clear that without the unifying notion of a plastic lens the Ridley invention could not have occurred.

21 For example, Binkhorst visited Ridley to understand the new methods, implanted copies of his lenses and then, dissatisfied with the size and weight of the early Ridley lens, he began a search for improvements.

22 Again, we find many competing routes to the solution of this problem. Well after Kelman succeeded with his method, others continued to search for a mechanical solution. In the 1970s, for example, surgeons at Moorfields reported on the use of a technique called lensectomy, in which the lens is cut mechanically and then aspirated. It was claimed to be cheaper than the phako technique and required less skill to perform although it could not be used with hard cataracts (Kanski and Crick 1977). It failed to catch on.

23 It is worth noting that the use of aspiration to remove cataracts long predated its application by Kelman but it was then restricted to the removal of soft cataracts, those typically experienced in patients under 30. Certainly, aspiration was well understood in the 1930s and Schere, in a paper in 1960, explained the use of a hollow needle and a hand syringe to remove a cataract while only creating a small incision. This method was widely used in the USA in the 1960s and its less invasive nature allowed rapid healing and the fitting of contact lenses to the younger patients (Rice 1967). The significance of the Kelman innovation is that it tackled the problem of hard cataracts, and thereby opened up a very large market for treatment.

24 They were manufactured by a company called Cavitron Surgical Systems, long since disappeared from the record.

25 The apparatus was soon improved by the incorporation of piezoelectric technology and ways of controlling the rate at which material is aspirated from the eye without causing major fluctuations in pressure.

26 The return of the posterior lens, located 'where nature intended', was in part also a consequence of improved lens design to which lighter materials and better methods of fixation were crucial. Shearing's J-loop lens, with a flexible haptic introduced in 1977, turned out to be the emergent dominant design and remains so to this day, although there have been a continuous stream of improvements in the materials and shape of the haptic element to better locate the lens.

27 Furthermore, in line with what Gelijns and Rosenberg (1999) have observed in relation to the medical device industry more generally, a noticeable geographic shift has taken place also in the domain of IOL-related devices. While the first two decades of the innovation are essentially a European story, the next three decades are told primarily in the United States with the involvement of major ophthalmic multinationals rising to dominate the industry. All of these firms have a major marketing and distributive presence in Europe but the preponderance of their innovation activity remains in the North American system.

28 Interestingly, these developments are based in materials science and the biotechnology of the eye and not upon engineering knowledge.

29 The shift from 'product' to 'process' is entirely in line with product cycle models of the innovation process. See, for example, Utterback (1994).

30 The relevant patents have been retrieved by the US Patent Office between May and December 2002 via targeted thematic searches. The patents have been fully inspected and reclassified by functions and by characteristics.

31 In the case of deformable lenses, the take off in activity is dated *circa* 1984. Between 1981 and 2002 a (cumulative) total of 74.8 per cent of the IOL patents are based on the use of deformable material.

32 This is a clear illustration of innovation as a 'distributed' process (Coombs *et al.* 2003).

33 At the Manchester Eye Hospital in 1991, for example, all the patients were inpatients, while by 1999 only 13 per cent fell into that category. As they are accurately described, most cataract patients are today ambulatory patients and this is a general phenomenon.

34 In 1980, this same Manchester hospital maintained *circa* 175 beds but, by the late 1990s, this number had fallen to 26.

35 For example, a large-scale survey of 100 ophthalmological hospital units in the UK, performing cataract surgery on some 18,000 patients over the period September 1997 to December 1998, found that 94 per cent of the lens insertions were in individuals over 65 (Desai *et al.* 1999a, 1999b).

36 An editorial in one ophthalmology journal declared: 'This has placed ophthalmology in the eye of a surgical storm'. A pioneer of the technique in the US commented: 'Even the most enthusiastic advocate of this procedure would agree that this has polarized the

American ophthalmic community like nothing else in recent memory' (both quoted in Jaffe 1999).

37 In 1978, the US Food and Drug Administration (FDA) initiated the largest clinical study on IOLs ever conducted. On 11 December 1981 this resulted in the Rayner-designed and manufactured Choyce MkVIII and MkIX lenses becoming the first IOLs to be approved by the FDA as safe and effective (Rayner 1999).

References

Apple, A. J. and Sims, J. (1996) 'Harold Ridley and the invention of the intraocular lens'. *Survey of Ophthalmology*, Vol. 40, pp. 279–292

Apple, D. J., Mamalis, N., Loftfield, K., Googe, J. M., Novak, L. C., Kavka-Van Norman, D., Brady, S. E. and Olson R. J. (1984) 'Complications of intraocular lenses: a historical and histopathological review'. *Surveys of Ophthalmology*, Vol. 29, pp. 1–54

Apple, D. J. *et al.* (2000) 'In tribute to Sir Harold Ridley'. *Survey of Ophalmology*, Vol. 45, Supplement, pp. S7-S12

Arnott, E. J. (1973) 'The ultrasound technique for cataract removal'. *Transactions of the Ophthalmological Society of the UK*, Vol. 93, pp. 33–38

Arnott, E. J. (1977) 'Kelman phaco-emulsification'. *Transactions of the Ophthalmological Society of the UK*, Vol. 97, pp. 60–63

Barraquer, J. (1956) 'The use of plastic lenses in the anterior chamber – indications-techniques-personal results'. *Transactions of Ophthalmological Society of the UK*, Vol. 74, pp. 537–552

Barraquer, J. (1959) 'Anterior chamber plastic lenses. Results of and conclusions from five years' experience'. *Transactions of the Ophthalmological Society of the UK*, Vol. 79, pp. 393–424

Binkhorst, C. D. (1959) 'Iris-supported artificial pseudophakia: a new development in intra-ocular artificial lens surgery'. *Transactions of the Ophthalmological Society of the UK*, Vol. 79, pp. 569–584

Choyce, D. P. (1960) 'The use of all-acrylic anterior chamber implants'. *Transactions of the Ophthalmological Society of the UK*, Vol. 80, pp. 201–219

Cleasby, G. W., Fung W. E. and Webster, R. G. Jr. (1974) 'The lens fragmentation and aspiration procedure (phacoemulsification)'. *American Journal of Ophthalmology*, Vol. 77, pp. 384–387

Coleman, A. and Morgenstern, H. (1997) 'Use of insurance claims to evaluate the outcomes of ophthalmic surgery'. *Surveys of Ophthalmology*, Vol. 42, pp. 271–278

Constant, E. W. (1980) *The Origins of the Turbo Jet Revolution*. Baltimore, MD: Johns Hopkins University Press

Coombs, R., Harvey, M. and Tether, B. (2003) 'Analysing distributed processes of provision and innovation'. *Industrial and Corporate Change* 12 (6), pp. 1125–1155

Desai, P., Reidy, A. and Minassian, D. C. (1999a) 'Profile of the patients presenting for cataract surgery in the UK: national data collection'. *British Journal of Ophalmology*, Vol. 83, pp. 893–896

Desai P., Minassianb, D. C. and Reidya, A. (1999b) 'National cataract surgery survey 1997–8: a report of the results of the clinical outcomes'. *British Journal of Ophthalmology*, Vol. 83, pp. 1336–40

Dowling, J. L. and Bahr, R. L. (1985) 'A survey of current cataract extraction techniques'. *American Journal of Ophthalmology*, Vol. 99, pp. 35–39

Duke-Elder, S. (1959) *System of Ophthalmology*. London: Henry Kimpton

Gelijns, A. and Rosenberg, N. (1999) 'Diagnostic devices: an analysis of comparative advantages' in D. Mowery and R. Nelson (Eds) *Sources of Industrial Leadership*. Cambridge, UK: Cambridge University Press

Hiles, D. A. and Hurite, F. G. (1973) 'Results of the first years' experience with phaco emulsification'. *American Journal of Ophthalmology*, Vol. 75, pp. 473–477

Jaffe, N. S. (1999) 'Thirty years of intraocular lens implantation: the way it was and the way it is'. *Journal of Cataract & Refractive Surgery*, Guest Editorial, 25 (4) April

Kanski, J. J. and Crick, M. P. (1977) 'Lensectomy'. *Transactions of the Ophthalmological Society of the UK*, Vol. 97, pp. 52–57

Kaufman, H. E. (1980) 'The correction of aphakia'. *American Journal of Ophthalmology*, Vol. 89, pp. 1–10

Kelman, C. D. (1973) 'Phaco emulsification and aspiration: a report of 500 cases'. *American Journal of Ophthalmology*, Vol. 75, pp. 764–768

Kelman, C. D. (1991) 'History of phacoemulsification' in M. D. Devine and W. Banko (Eds) *Phacoemulsification Surgery*. London: Pergamon Press

Leaming, D. V. (1998) 'Practice styles and preferences of ASCRS members'. *Journal of Refractive and Cataract Surgery*, Vol. 24, pp. 552–561

Linebarger E. L., Hardten, D. R., Shah, G. K. and Lindstrom, R. L. (1999) 'Phacoemulsification and modern cataract surgery'. *Surveys of Ophthalmology*, Vol. 44, pp. 123–147

McCarty, C. A., Keeffe, J. E. and Taylor, H. R. (1999) 'The need for cataract surgery: projections based on lens opacity, visual acuity, and personal concern'. *British Journal of Ophthalmology*, Vol. 83, pp. 62–5

Nørregaard, J. C., Bernth-Petersen, P. and Andersen, T. F. (1996) 'Changing threshold for cataract surgery in Denmark between 1980 and 1992'. *Acta Ophalmologica Scandinavica*, Vol. 74, pp. 604–608

Nørregaard, J. C., Bernth-Petersen, P., Alonso, J., Dunn, E., Black, C., Andersen, T. F., Espallargues, M., Bellan, L. and Anderson, G. F. (1998) 'Variations in indications for cataract surgery in the United States, Denmark, Canada, and Spain: results from the International Cataract Surgery Outcomes Study'. *British Journal of Ophthalmology*, Vol. 82, pp. 1107–1111

Patel, A., Carson, D. R. and Patel, P. (1999) 'Evaluation of an unused 1952 Ridley intraocular lens'. *J. Cataract and Refractive Surgery*, Vol. 25, pp. 1535–1539

Rayner (1999) Rayner 50th Anniversary 1949–1999, Rayner Intraocular Lenses: Hove.

Rice, C.S. (1967) 'Lens aspiration: a method of treatment for soft cataracts'. *Transactions of the Ophthalmological Society of the UK*, Vol. 87, pp. 491–498

Ridley, H. (1946) 'Recent developments in the manufacture, fitting and prescription of contact lenses of regular shape'. *Proceedings of the Royal Society of Medicine*, Vol. 39, pp. 467–499

Ridley, H. (1951) 'Intra-ocular acrylic lenses'. *Transactions of the Ophthalmological Society of the UK*, Vol. 71, pp. 617–621

Ridley, H. (1964) 'Intraocular acrylic lenses: past, present and future'. *Transactions of the Ophthalmological Society of the UK*, April, pp. 5–14

Schere, H. G. (1960) 'Aspiration of congenital or soft cataracts: a new technique'. *American Journal of Ophthalmology*, Vol. 50, pp. 1048–1056

Schlote, T., Sobottka, B., Kreutzer, B., Thiel, H. J. and Rohrbach, J. M. (1997) 'Cataract surgery at the end of the 19th century'. *Surveys of Ophthalmology*, Vol. 42, pp. 190–194

Sen, A. (1999) *Development as Freedom*. Oxford, UK: Oxford University Press

Stark, W. J., Terry, A. C., Werthen, D. and Murray, G. C. (1984) 'Update of intraocular lenses implanted in the United States'. *American Journal of Ophthalmology*, Vol. 98, pp. 238–239

Usher, A. P. (1929; 1954) *A History of Mechanical Inventions*. New York: Dover Publications

Utterback, J. M. (1994) *Mastering the Dynamics of Innovation*. Boston, MA: Harvard Business School Press

West, S. and Valmadrel, C. (1995) 'Epidemiology of risk factors for age related cataract'. *Surveys of Ophthalmology*, Vol. 39, pp. 323–334

Wybar, K. (1984) *Ophthalmology*. London: Baillière Tindall

2 Coronary artery disease

Andrea Mina, Ronnie Ramlogan,
Stan Metcalfe and Gindo Tampubolon

Introduction

Cardiovascular diseases are a leading global cause of death (WHO 2011). Coronary artery disease (CAD) is the most common type of cardiovascular disease: it is the end result of a process called atherosclerosis, by which an atheroma – a plaque of fatty deposits – forms on the inner layer of the coronary artery and impedes the flow of blood to the heart. In the early stages, the build-up of these deposits is silent (symptom-less) but, as the disease progresses, it can induce a variety of conditions ranging from mild chest pains (angina) and short-ness of breath to heart failure and sudden death.

In this chapter we investigate what has been regarded by many as one of the most important medical breakthroughs of the last decades of the twentieth century.[1,2] Percutaneous Coronary Intervention (PCI) or, as it is more familiarly known, coronary angioplasty[3], emerged in the late 1970s to become a technique that is now more commonly used than the principal surgical alternative, coronary artery bypass surgery (CABG), in the treatment of advanced coronary artery disease. With the benefit of experience, PCI has proved to be an effective mode of treatment that has also brought with it a major transformation in the division of labour in cardiology, with the field of Interventional Cardiology being recog-nised as a separate and distinctive sub-speciality of the broader field of practice. But it has also proved to be a black box of uncertainty in which progress is hesi-tant in the face of challenging emergent medical problems. Mina *et al.* (2007) and Ramlogan *et al.* (2007) present large-scale bibliometric analyses of the medical literature on PCA and of patent records of relevant inventions. In this chapter we revisit the way in which this innovation emerged as a set of problem-solving sequences and identify the mechanisms of operations of a 30-year long co-evolu-tionary process of knowledge, technology and institutional change.

Treatments for CAD prior to the 1980s

For sufferers of the disease chest pains, angina, can be induced by physical exer-tion and/or emotional stress. These increase the heart's requirement for oxygen-enriched blood but, with narrowed arteries restricting blood supply, the sufferer experiences severe discomfort. The pain usually becomes evident when the vessel is only able to deliver at 30 per cent of its capacity. At this level of closure, the oxygen-enriched blood that the heart receives is only adequate when the body is at rest. A heart attack may occur when a coronary artery becomes completely blocked and insufficient oxygen supply to the heart, called ischemia, results in the death of the heart muscle.

As recently as the 1960s, treatment options for angina or acute myocardial infarction (heart attack) consisted of few medications (mainly nitroglycerin), rest and hope.[4] Between the 1960s and 1970s beta blockers and calcium channel blockers were added to the cardiologist's arsenal for dealing with angina. Unfortunately, while these started to provide effective relief by reducing the frequency and force of the heartbeat, they were not a cure for the underlying problem. Surgical treatment options also improved from the 1960s with the development of coronary artery bypass surgery.[5] This is a major invasive and complicated surgical procedure taking between three and six hours to perform. It requires general anaesthesia, the use of a heart-lung machine to substitute for heart-lung functioning during surgery and a lengthy post-surgery recuperation period. Put simply, the idea behind the procedure is to improve the blood flow to the heart muscle by bypassing the blockages with blood vessels harvested from the leg (saphenous vein) or chest wall (internal mammary artery).

At the time of its introduction bypass surgery was regarded as a revolutionary procedure. The idea of stopping a heart, restoring its blood supply then restarting it, bordered on the miraculous. The technique spread rapidly although figures for the volume of procedures undertaken in the early period are patchy. In 1973 around 25,000 operations were performed in the US and this increased to 70,000 by 1977 (OTA 1978). Elsewhere, the absolute number of procedures was small in comparison. In the UK, for example, 2,297 operations were carried out in 1977 and this increased to 4,057 by 1980 (British Heart Foundation 2004).

The diffusion of this procedure was not without controversy and this was primarily related to the evidence base on which bypass surgery was being promoted. A debate raged throughout the 1970s about the quality of evidence that was being assembled about the efficacy of bypass surgery relative to medical therapy.[6] So much so that the US Office for Technology Assessment (OTA 1978) was decidedly lukewarm about the procedure arguing that, in terms of survival, the VA randomised trial and a number of others showed that surgery did not appear to bring any appreciable longevity benefit when compared with medically treated patients. The OTA did concede the point that bypass surgery gave 'excellent symptomatic relief from angina pectoris' (p. 43) although it was careful to caution about placebo effects associated with surgery.[7]

The advent of a new treatment modality

It is against this background of uncertainty about the efficacy of coronary bypass surgery that Percutaneous Coronary Intervention (PCI) emerged. Its development was truly original but it did nevertheless build on prior knowledge and existing techniques, as is often the case in the history of medical technologies. The two techniques that laid the foundations for PCI were cardiac catherisation and transluminal angioplasty.

Cardiac catheterisation is a diagnostic procedure in which a catheter (a thin flexible tube) is inserted into the right or left side of the heart. This procedure could then be used to produce angiograms (x-ray images) of the coronary arteries and the left ventricle, the heart's main pumping chamber, and/or used to measure pressures in the pulmonary artery and to monitor heart function. Werner Forssmann is credited with being the first to introduce a (urological) catheter into the right atrium (his own!) in 1929. Branded as 'crazy' by his contemporaries, his immediate reward for this achievement was dismissal from his German hospital (he went on to win the Nobel Prize in 1956). By the 1950s, however, following the work of Cournand, Seldinger and others, diagnostic catheterisation had become established as the main technique for investigating cardiac function.

The American Charles Dotter coined the term Percutaneous Transluminal Angioplasty (PTA) (Rosch *et al.* 2003) for a treatment that he developed in the 1960s. During an angiographic examination he inadvertently reopened an occluded right iliac artery (in the lower abdomen) by pushing through it with an angiography catheter. Realising the therapeutic potential of dilating narrowed arteries, Dotter went on to refine the technique but, as was the case with Forssmann, this was met with scepticism by the American medical fraternity. However, European radiologists took a more positive view and soon institutionalised the term 'dottering' (of patients) to refer to improving the patency of arteries by the introduction of a series of coaxial catheters. One of Dotter's European followers, Eberhardt Zeitler, used the technique on a large number of patients and this helped the diffusion of the technique among practitioners.[8] Among them was the German radiologist Andreas Gruentzig, who is named in the annals of medical history as the pioneer of PCI.

Gruentzig was exposed to the Dotter method when he was based at the Ratchow Clinic in Darmstadt (Germany) in the mid-1960s. At Ratchow he was a Research Fellow in Epidemiology studying coronary artery disease but later, wanting to become a cardiologist, he moved to the University of Zürich (in 1974). Here he began to think about whether the Dotter method could be applied to the heart recognising that 'any application of the dilatation procedure to other areas of the body would require technical changes' (King 1996, p. 1624).

Encouraged by his colleague and Joint Head of Cardiology, Wilhelm Ruttishauser, Gruentzig proceeded to develop by himself new prototypes of catheters with a soft balloon at the tip that could be inflated to break the plaque inside the vessel, the foundation of what he called Percutaneous Transluminal Coronary Angioplasty (PTCA). Two crucial developments followed. First, in 1972, he introduced a balloon made of PVC, a tough, less compliant material than latex, with which he had experimented earlier. Second, in 1975, he developed a new single and then, more importantly, a double lumen catheter. This was a single catheter with two tubular channels, one for inflating the balloon and the other for injecting contrast media and monitoring intravascular pressure. Some early results were presented to the American Heart Association meeting in 1976 to a largely sceptical audience but, by 1977, he had succeeded in performing the first PTCA on a human patient in Zürich. The technique diffused relatively quickly thereafter, particularly in the US.

The development of PCI: phase 1

The development of PCI can be broken down into several overlapping phases of activity involving both clinical practice and technology. The first phase was connected with the development of PTCA or balloon angioplasty, Gruentzig's original breakthrough. This was a phase of intense exploration of the search space opened up by this new technology. Practitioners were concerned with identifying the medical conditions under which this new technique would deliver benefits to patients under tolerable margins of risk. A study by Cowley *et al*. (1985), for example, produced under the sponsorship of the National Heart Lung and Blood Institute (NHLBI), provided early evidence of the effectiveness of balloon angioplasty procedures in medical centres across the US and in other countries. This was one of a number of foundation papers that provided early evidence from a Registry set up by NHLBI in 1979 to collect, analyse and disseminate the results from using the balloon angioplasty procedure in medical centres across the US and in other countries (Mullins *et al*. 1984). By the year 2000 NHLBI had set up several registries to understand the long-term efficacy of PTCA and the alternatives approaches that were emerging. The first registry followed 3,079 patients who received PTCA between 1977 and 1982. A second registry followed the outcomes of 1,500 patients from the first registry for a minimum of five years plus an additional 2,000 newly entered patients who received PTCA in 1985 and 1986, while a third registry, the New Approaches to Coronary Intervention (NACI), followed approximately 4,424 patients between November 1990 and February 1997.[9]

Two major unexpected medical problems, acute occlusion and restenosis, emerged to trouble the early diffusion of PTCA.

Acute occlusion

The PTCA procedure involves making small incisions in the groin or arm under local anaesthetic where a catheter can be fed through to the obstructed coronary vessel. The balloon attached to the end of the catheter is used to unblock the vessel and consequently restore the blood flow. In a small but significant number of cases, the procedure resulted in weakening and collapse of the internal structure of the artery which would subsequently require emergency bypass surgery. Several studies in the early 1990s comparing the outcomes of CABG with PTCA showed higher re-intervention rates with PTCA (Reul 2005). The incidence of acute vessel closure was 3–5 per cent, and would occur within the first 24 hours of the procedure due to vessel dissection or acute thrombus formation (Hamid and Coltart 2007). Thus, in the early days, the practise of PTCA was contingent on having emergency coronary artery by-pass operating facilities available.

Second, tissue trauma at the site of the procedure sometimes triggered blood clotting which, depending on severity, would require major invasive treatment. Over time, however, the occurrence of this would be countered with anti-thrombolytic drugs.

Restenosis

The second major problem was restenosis – the appearance of a new constriction in the artery (Pepine *et al.* 1990). It had a high incidence rate, between 25–50 per cent in patients having undergone balloon angioplasty. This constriction was not atherosclerotic in nature but resulted from the outgrowth of 'endothelial' cells that normally line blood vessels and has been likened to 'over-exuberant' tissue healing and regeneration similar to scar formation. This tended to occur during the first three to six months after the procedure and effectively nullified the intervention so that the patient would require further revascularisation.

Various candidate devices were tested to solve the problems but these were of limited success in dealing with restenosis. For example, in contrast to dilating and compressing the plaque John Simpson developed an atherectomy device to remove it with barotrauma (change in pressure) (King 1998). Such attempts eventually led to the development of the rotational atherectomy device, a diamond-coated burr rotating at around 150,000 revolutions per minute to remove plaque through abrasion. While this device met with limited success, high complication rates, cost and the paucity of data from randomised trials, limited widespread acceptance and this technique has been largely confined to patients with hard, calcified plaque (Whitlow 1997). Another method, laser angioplasty, used the energy created by an argon laser focused on the end of a catheter to ablate tissue on contact with the catheter tip, but this did not compare favourably with balloon angioplasty (King 1998).

Phase 2: the introduction of the stent

The second phase in the development of the new procedure witnessed the introduction and development of the stent in response to the problems that were limiting the potential of the new technique. The term 'stent' originated in dentistry and dates back to the nineteenth century, but it appeared in the non-dentistry medical literature when used by Dotter in 1983 in a study of percutaneously implanted vascular endoprosthesis in canine experiments (Balcon *et al.* 1997). In the late 1980s stent developed into a metal meshwork structure (Ruygrok and Serruys 1996) which could be inserted with the balloon catheter, opened up and left in the vessel to support the walls of the artery and prevent acute vessel closure and restenosis.

Ulrich Sigwart performed the first coronary stent procedure in 1984 in Lausanne (Towers and Davies 2000) and several papers published between 1985 and 1988 described the progress with the balloon-mounted coronary stent developed by Julio Palmaz and Richard Schatz (and produced by Johnson & Johnson)

(Kent 2010), in both canines and humans, and the self-expanding Wall stent used in humans.

The improved technique was met favourably by practitioners and rapidly diffused. By 1999, coronary stenting was performed in 84.2 per cent of PCI procedures (Serruys *et al.* 2006) but, as with PTCA before it, the diffusion of stents into wider practise presented the cardiology community with unanticipated new challenges. In many cases the use of a stent resulted in a sub-acute thrombotic coronary artery. This is an occlusion (blood clot) at the stented site that formed anywhere between 24 hours and up to 30 days after the procedure. The possibility of such an event necessitated a complex anti-coagulation treatment and prolonged hospital stays. However a further unexpected complication arose, that of in-stent restenosis. Under normal circumstances new tissue grows inside the stent covering the metal mesh when this is placed inside the artery. This allows the smooth flow of blood over the stented area. Subsequently, however, scar tissue may develop underneath the new healthy lining and this may be so thick that it obstructs the blood flow. This new kind of restenosis usually occurs between 3 to 6 months after the procedure and affects about 25 per cent of patients (Dangas and Kuepper 2002).

The enthusiasm for stents waned considerable in the early 1990s, particularly as the evidence being gathered showed that stenting did not give a better success (or lower complication) rates than those of routine balloon angioplasty (Balcon *et al.* 1997). However, the results from three important trials marked a turning point for the medical community and favoured increasing acceptance and use of stenting technology as an integral part of PCI. Studies by Serruys *et al.* and Fischman *et al.* in 1994 followed by Colombo *et al.* (1995) provided persuasive evidence of the advantages of stenting compared with 'simple' balloon angioplasty and to the finding that the success rate of the procedure depended heavily on the placement of the stent, focusing on full expansion, adequate deployment of the stent using intravascular ultrasound and by the use of simplified and more effective anticoagulation protocols.

Subsequently, further exploration focused on improving the use of stents and dealing with the problem of the plaque reforming inside the stent (in-stent restenosis). This very problem also triggered the emergence of parallel trajectories of research, principal among them being drug-eluting stents in which stents coated with drugs are locally delivered to the point of the lesion.

While the introduction of stenting technology reduced the restenosis rate from 40 percent to around 30 per cent (Fischman *et al.* 1994), it created a new problem - 'in-stent restenosis' - and drug-eluting stents (DES) were developed to specifically target this problem. Stents were now coated with a polymer that released a drug to inhibit the cell proliferation causing restenosis over the course of 6 to 9 months. Cordis launched its Cypher® stent in 2003, and this was followed by the Taxus® stent from Boston Scientific in 2004. In subsequent years other manufacturers followed with variations in the design (type of metal used; strut thickness; mechanics of strut interlinkage), drug used and the release characteristics of the polymer.

DES quickly attained a high penetration, accounting for around 50 per cent of the coronary stent market (Li and Kozlek 2012) due to the lower rates of restenosis, although there was an initial lack of long-term safety and efficacy data. Subsequent studies compared individual DES both with their bare metal (BMS) equivalents and with each other (see, for example, Mahmoudi *et al.* 2011). In general the evidence has shown that, compared with BMS, DES significantly reduce the incidence of restenosis to levels of under 10 per cent (Farooq *et al.* 2011) but meta-analyses were beginning to show that DES had a much greater risk of very late stent thrombosis (where a blood clot forms inside the stent more than a year after insertion) compared with BMS (Bavry *et al.* 2006). Thus the consensus that emerged was that DES require a longer period of dual antiplatelet therapy relative to BMS in order to prevent stent thrombosis. On-going studies continue to evaluate newer stent platforms, drugs, polymers, polymer-free stents and bio-absorbable stents and it is becoming clear that there will not be one single stent suitable for all patients and lesions (Wilson and Cruden 2013).

Beyond technology: co-evolutionary dynamics of knowledge and institutions

The development of coronary angioplasty is a story of entrepreneurial individuals as well as the building of a community of practitioners. It unfolds along interrelated epistemic, social and institutional dimensions. In spite of the initial conservative reaction to his work, Gruentzig's pioneering insights and first successes opened up tremendous opportunities for further improvements of the new technique whereby a growing number of practitioners started experimenting with it and a growing range of devices started to be developed after the introduction of the first tested prototypes (Figure 2.1).

In this respect, the nature of this innovation process mirrors that of other significant medical innovations in being an uncertain co-evolutionary process of knowledge and technique (Gelijns and Rosenberg 1994; Gelijns *et al.* 1998; Gelijns *et al.* 2001) in which trajectories often emerge in the form of sequences of innovative ideas (Metcalfe *et al.* 2005). These involve coherent directions of change – or trajectories (Dosi 1982) – and signal the cumulativeness of interdependent research activities whose results build on previous knowledge. Furthermore, these imply specific configurations (for example, technical designs) embodying ways of combining knowledge that gradually become formal or informal standards (Utterback 1994) in the unfolding search for solutions to problems.

The growth of PCI practice was accompanied by many improvements in devices and practice including the invention of the steerable balloon catheter by Simpson in the early 1980s (Simpson *et al.* 1982). It is this contribution that enabled clinicians to access the most distal lesions and to achieve greater direction control of the catheter through the coronary system. Most importantly, the structure of these many complementary contributions to the innovation sequence also reflects the shift in the nature of the dominant problem over time. The solution to the catheter problem and Greuntzig's balloon device to compress the

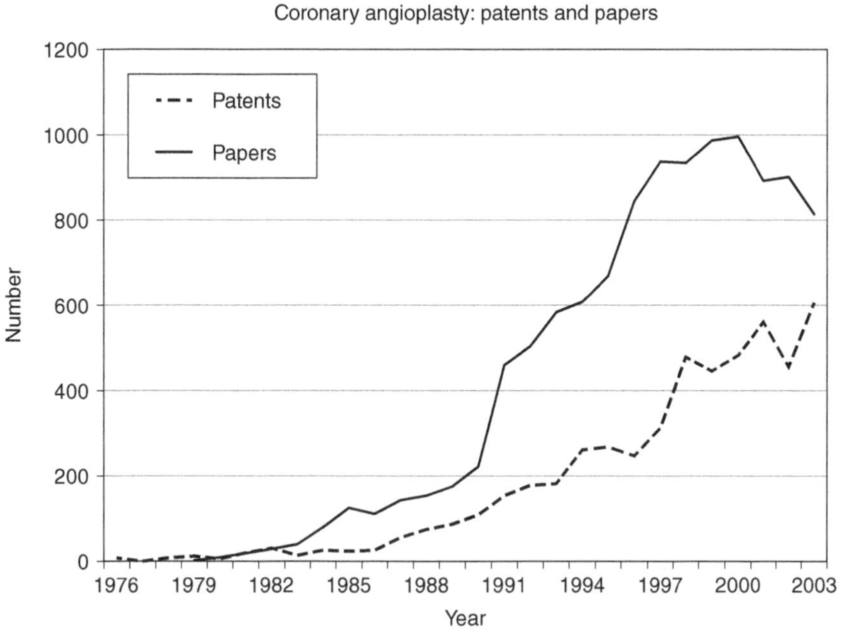

Coronary angioplasty: patents and papers

Figure 2.1 The growth of codified knowledge.

Source: Mina *et al*. 2007, Fig. 3, p.797, with permission.

plaque opened up new territory but it was soon found that restenosis – the refor-mation of the plaque after the procedure – occurred in a significant number of patients, drastically reducing the efficacy of the treatment and raising its real cost. The solution to this problem was the invention and innovation of the stent (Eeckhurst *et al*. 1996).

As we have observed above, after the original breakthrough further exploration focused on improving the use of stents and dealing with the problem of the plaque reforming inside the stent (in-stent restenosis). This very problem also triggered the emergence of parallel trajectories of research. Among the various solutions that were explored were stents coated with drugs that are locally delivered to the point of the lesion, pharmacological therapies that precede or follow stenting, and radiation therapies. And the search for yet better solutions is ongoing.

An important aspect of this process was the initial conditions for the early diffusion of PCI. When Gruentzig presented his first results to the American Heart Association conference in 1977 his intervention generated enough interest for him to receive numerous requests from other cardiologist wanting to learn the technique. A close-knit – and practice-based – micro community developed around Gruentzig that constituted the backbone for the diffusion of the new tech-nique.[10] It included, among others, Richard Myler who, together with Simon

Sterzer, was the first to perform angioplasty in the United States in 1978 (King 1998). Over time Gruentzig's connections with the US intensified to the point that in 1980 he moved from Zurich to Emory University, where he had fewer constraints on the lab time he was allowed by his employer. At Emory not only could he work on an increasing number of cases but he was also provided good teaching facilities, which he was very keen to have. He was convinced that a critical mass of expert practitioners needed to be trained as well as possible to avoid the formation of imperfect skills that would have jeopardised the long-term success of the new technique.

It is in this community that we find the seeds of science as well as the seeds of the industrial complex that was to grow over the following three decades. Gruentzig was not only scientifically minded but also entrepreneurial. His prototype catheter was developed, of all places, in his kitchen and he later entered into a relationship with Schneider, a Swiss medical needle manufacturing company based in Zurich, for manufacturing a marketable device and he applied for a patent with the US Patent and Trademark Office in 1977. This was granted in 1980 and paved the way for the diffusion of coronary angioplasty in the US.

Gruentzig also exerted a considerable degree of control over the production and the commercialisation of his catheters.[11] In the early days in Zurich he exerted tight control of the sale of the first angioplasty catheters that were being produced under his instructions by Schneider, a small Swiss company later to be acquired by Pfizer in 1984 and sold on to Boston Scientific in 1998. It has been reported that he required practitioners who bought catheters to receive training or counselling from him on how best to use the device.[12] This significantly contributed to a reduction in the rate of failure of the procedure due to improper use in a phase of the diffusion process that was most delicate for the future success of angioplasty.

When the procedure started to be experimented with on a larger scale, the need also emerged to evaluate the performance of the technique in a systematic way. John Abele, who went on to found one of today's market leaders (Boston Scientific) and who started collaborating with Gruentzig in the development of double-lumen catheters, recalls that there were informal discussions between Gruentzig and Myler and himself about creating a registry to document what was being learnt in clinical practice wherever the technique was used. They then developed a very simple but comprehensive form where results of every procedure could be recorded. The registry that resulted from the collection and organisation of the forms lasted for no longer than 18 months but was not to disappear. In 1979 the National Heart, Lungs and Blood Institute (NHLBI) – the relevant branch of the National Institute of Health – invited a small working group to discuss how evaluation of PTCA could be managed. It was agreed that a workshop should be held to review the preliminary evidence and that a voluntary registry be set up. The PTCA registry, which we mentioned earlier, was established in March 1979 (Mullins *et al*. 1984) and was really a formalisation of the earlier database started by Gruentzig and others.

Thus, in this process of division of social labour, the monitoring function of the system passed to the NHLBI while the regulation of medical devices shifted to the FDA in the same years. Both these complementary modes of governance proved crucial to the extension of understanding and practice, and to the growth of the market on which investment by commercial firms depended. Market institutions do not usually, if ever, operate independently of wider regulatory norms and these norms – formal and informal – like the market itself, often are created and co-evolve with it. This case is a good illustration of this point.

Prior to the 1980s, as we have mentioned, Gruentzig and Myler personally exerted informal control of the availability, quality and safety of catheters. In a way, they acted as the first regulators of angioplasty devices before the FDA started implementing the reform of Medical Devices Law passed in 1976 and became the ultimate arbiter for regulation of medical technologies. The FDA then became the institution that, in the public interest, exerted functions of evaluation and selection of the variety of drugs and devices stemming from the wealth of R&D activities that followed the sparse experimentation of the early years. Needless to say, its role grew in importance along with, on the one hand, the increasing commercial interests associated with angioplasty, which grew fast especially in the US, where the centre of gravity of the community shifted both scientifically and technologically.

Also, as the technique developed, a number of complementary and competing solutions were found in the community gathered around Gruentzig. From there, these new solutions found their way to the market very often through the activities of a number of physician-entrepreneurs such as John Simpson, inventor of the steerable guidewire, who founded no fewer than ten product-innovation based start-ups and then sold them in rapidly expanding knowledge markets to larger device manufactures (Guidant, Abbott Laboratories and Boston Scientific among others). The early development of stents is a case in point. A patent analysis by Xu *et al.* (2012) over the period 1984–1994 showed that small privately held companies created by physician inventors contributed the most patents as well as the most highly cited patents. The ten most highly cited patents in the field were dominated by two private firms, Expandable Grafts Partnership associated with Julio Palmaz and Richard Schatz (the Palmaz-Schatz stent that was licensed to Johnson & Johnson), and Cook Inc. who had an established relationship with pioneering interventional radiologist Cesare Gianturco (the Gianturco-Roubin Flex-Stent) (Xu *et al.* 2012). Both of these were the first stents to be approved by the FDA.

If we consider the overall evolution of the market for PCI devices, we find again not simply a process of growth but a process of punctuated qualitative change in the nature of the problems that are addressed and in the new ways found to solve them. Figure 2.2 proxies the dynamics by looking at the FDA device-approval records, which give a very good picture of the market output of the broader technological exploration that has been charted through extensive networks analysis in previous work (Mina *et al.* 2007; Ramlogan *et al.* 2007). The figure shows the (percentage) composition of FDA-approved products by type of innovations. It illustrates, first of all, the relative 'closure' of the technical problem solved by coronary balloon catheters, which over time accounts for

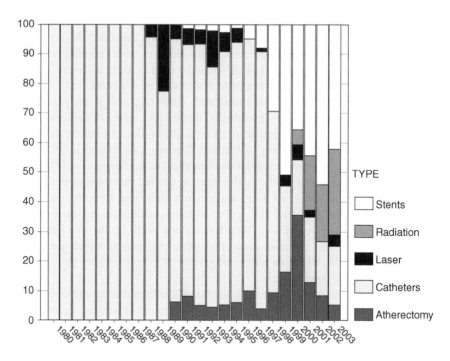

Figure 2.2 Percentage composition of FDA market approvals.

Source: Mina 2009, Fig. 8, p. 459.

proportionally less and less inventive efforts in favour of other types of catheters and stent-related products. Secondly, it documents the rise of devices that address either complementary or subsequent challenges that arose through practice along the development of the problems-solutions sequence (or dominant technological trajectory) associated with the surgical treatment of the disease.

Most interestingly, the transformation over time of the scientific and techno-logical knowledge bases is deeply connected with the industrial transformation of the PCI medical device segment. In the earlier stages the industry focussed on the catheter and stent areas, where dominant designs soon emerged. Subsequently, new windows of opportunities opened up along the unfolding technological trajectory: companies from outside the established supply chain with radically different competence bases (for example in the areas of radiation or laser technol-ogy) started to challenge the market leaders and began to account for an expand-ing proportion of PCI devices. The resulting picture is, on the one hand, one of a Schumpeterian process of competition through which incumbents are continu-ously challenged by innovators and, on the other, a highly distributed innovation system where some firms specialise in complementary market segments while others compete for alternative solutions along the problem sequence.

Conclusion

The development of PCI is a story of how the creative vision of a few lead developers and users became shared by an increasing number of practitioners and spilled over, on the one hand on the need for market and non-market co-ordination mechanisms and, on the other, on the formation of growing profit expectations. Insofar as the advancement of codified knowledge is concerned there is a close correspondence between the development of scientific understanding of the disease and the technological tools used to diagnose and treat the conditions, which is a clear mark of the strong complementarities (but not of linear dependency) between basic and applied medical research. Furthermore, as science and technology co-evolved, so did the set of interdependent problems that emerged along the search for satisfying cures to the disease because, as one problem was solved, others presented themselves, partly in reaction to the newly found solutions, and were opportunities for further change in the nature and composition of the product market.

Certainly a great number of factors contributed to the emergence and diffusion of coronary angioplasty jointly with the exceptional skills of the pioneers, the unpredictable consequences of early successes and the organisation, expansion and renewal of the supply capacity of the sector (Mina *et al.* 2007). At least two other factors were of the utmost importance in enabling the development of minimally invasive therapeutics for coronary artery disease: the concomitant growth of the US venture capital market and the effective institutionalisation of demand for coronary angioplasty. The lion's share of technological breakthroughs was launched since the early 1980s by VC-backed companies and the phenomenon of serial entrepreneurship is a constant pattern of innovation in the field. Overall, it has been estimated that over the past 20 years the contribution of venture capital to the sector amounted to almost 15 billion dollars in more than 2,000 deals for the development of minimally invasive treatments for cardiovascular disease.

With respect to the market need, the order of magnitude of the problem is self-explanatory: heart diseases are the leading cause of death in Western economies and also the costliest among all major preventable chronic diseases. As far as the early introduction of PTCA is concerned, the new technique offered a less invasive and therefore less risky alternative to the more established (but by no means uncontroversial) coronary bypass graft surgery (CABG). However, what initially appeared as a cost-effective measure did not prove so cost-effective once, through practice, it became clear that it was associated with the problem of restenosis. This meant iterations of the procedure, which in turn entailed additional costs and the consequent erosion of the productivity improvement of PTCA. It was only through further improvement through learning and additional technical change, above all the introduction of stents, that the cost-effectiveness of the procedure dramatically improved. This could not possibly have been foreseen at the time of introduction of PTCA. Furthermore, by gradually improving performance, over time incremental technical change radically transformed PTCA from a complement to CABG to a substitute for it (Cutler and Huckman 2003). Again, this could not have been foreseen at the time of early diffusion of PTCA. Needless to say, the consequences of this are far reaching for economic analysis because they question the value of short term cost-benefit analyses and the opportunity costs of neglecting the fundamental uncertainty associated in the early stages with what have become, in the long run, star performers of the health technology sector.

Overall, medical progress in this clinical area takes the form of a series of unfolding trajectories of change that converged and diverged along specific sequences of problems. These can only be appreciated *ex post* as the result of a process of path-dependent emergence through multi-level mechanisms of selection weeding out over time knowledge that is judged as ineffective through clinical use by the relevant communities of practitioners, who typically operate across the boundaries between science, technology and clinical practice, especially in early phases of development of new treatments.[13] We have also found that the sequence of scientific and technological problems solved along the main trajectory of development, itself an outcome of the dispersed and coordinated interaction of different agents and institutions, bears important implications for the structure and the composition of the market: this not only grew over time, but was also crucially and continuously transformed by new knowledge accumulated over time in the fight against the disease through a co-evolutionary process of epistemic and institutional change.

Acknowledgements

Figure 2.1 reprinted from Mina, A., Ramlogan, R., Tampubolon, G. and Metcalfe, J. S. (2007) 'Mapping evolutionary trajectories: applications to the growth and transformation of medical knowledge', *Research Policy*, Vol 36, pp. 789–806. Copyright 2007, reproduced with permission from Elsevier.

Figure 2.2 reprinted from Mina, A. (2009) 'The emergence of new knowledge, market evolution and the dynamics of micro-innovation systems', *Economics of Innovation and New Technology* 18(5): 447–466, Figure 8, p. 459. Copyright 2009. Reprinted by permission of Taylor & Francis Ltd, http://www.tandfonline.com.

Notes

1 We would like to extend our deep appreciation and thanks to: Dr Luigi Venetucci and Professor Anthony Heagarty (Manchester Medical School); Dr Bernard Clarke (Manchester Royal Infirmary); Professor Spencer B. King III (Fuqua Heart Center, Atlanta); Martin B. Leon MD and Eric A. Rose MD (Columbia Presbyterian Hospital, New York); Professor Annetine Gelijns (and her colleagues at the International Center for Health Outcomes and Innovation Research, Columbia University, New York). They have all been extremely generous with their time, advice and support and have has helped to sharpen our understanding of the development of coronary angioplasty.

2 A survey of general internists actively involved in patient care by Victor Fuchs and Harold Sox in 2001 ranked coronary angioplasty 3rd (only behind MRI and CT scanning and ACE inhibitors) of the 30 most important medical innovations over the last 25 years.

3 The earlier literature refers to this technique as Percutaneous Transluminal Coronary Angioplasty (PTCA).

4 The use of nitrates (and nitroglycerin) dated back to the mid- to late-1800s. These provided transient relief by dilating vein and arteries thereby enabling more blood to get to the heart.
5 See, for example, Connolly (2002) for an account of the development of bypass surgery.
6 The Mullins and Lipscomb (1977) review of the literature pointed to incomplete analyses and less than ideally designed studies as few randomised studies were available at that time.
7 One of the interesting features of the spread of bypass surgery is how quickly it was taken up for cohorts of patients for which the medical evidence was not indicated. Anderson and Lomas (1988) expressed concern about the apparent change in clinical policies towards the use of bypass procedures in the elderly without solid evidence on efficacy or cost effectiveness.
8 This account draws in part on King (1996, 1998) and authors' original interviews with him on 10 May, 2005.
9 See http://clinicaltrials.gov/ct2/show/study/NCT00005677
10 See also Interview with John Abele (Cohen and Klepper 1996), Part III, available at www.ptca.org/nv/interviewframe.html
11 Interview with Heliane Canepa (Cohen and Klepper 1996), then President of Schneider Worldwide and former collaborator of Gruentzig. Available online at: www.ptca.org/nv/interviewframe.html
12 Interview with John Abele (Cohen and Klepper 1996). Available online at: www.ptca.org/nv/interviewframe.html
13 This does not imply that the outcomes are necessarily optimal or could not have been different!

References

Anderson, G. M. and Lomas, J. (1988) 'Monitoring the diffusion of a technology: coronary bypass surgery in Ontario'. *American Journal of Public Health*, Vol. 78, pp. 251–254

Balcon, R., Beyar, R., Chierchia, S., De Scheerder, I., Hugenholtz, P. G., Kiemeneij, F., Meier, B., Meyer, J., Monassier, J. P. and Wijns, W. (1997) 'Recommendations on stent manufacture, implantation and utilization'. *European Heart Journal* 18, pp. 1536–1547

Bavry, A. A., Kumbhani, D. J., Helton, T. J., Borek, P. P., Mood, G. R. and Bhatt, D. L. (2006) 'Late thrombosis of drug-eluting stents: a meta-analysis of randomized clinical trials'. *Am J Med*. 119, pp. 1056–1061

British Heart Foundation (2004) 'Coronary Heart Disease Statistics'. Available online at http://www.heartstats.org/uploads/documents%5C2004pdf.pdf

Cohen, W. and Klepper, S. (1996) 'Firm size and the nature of innovation within industries'. *Review of Economics and Statistics* 78(2), pp. 232–243

Colombo, A., Hall, P., Nakamura, S., Almagor, Y., Maiello, L., Martini, G., Gaglione, A., Goldberg, S. L. and Tobis, J. M. (1995) 'Intracoronary stenting without anticoagulation accomplished with intravascular ultrasound guidance'. *Circulation* 91, pp. 1676–88

Connolly, J. E. (2002) 'The development of coronary artery surgery: personal recollections'. *Texas Heart Institute Journal*, 29(1), pp. 10–14

Cowley, M. J., Mullin, S. M., Kelsey, S. F., Kent, K. M., Gruentzig, A. R., Detre, K. M. and Passamani, E. R. (1985) 'Sex differences in early and long-term results of coronary angioplasty in the NHLBI PTCA Registry'. *Circulation*, 71, pp. 90–97

Cutler, D. M. and Huckman, R. S. (2003) 'Technological development and medical productivity: the diffusion of angioplasty in New York state'. *Journal of Health Economics*, 22(2), pp. 187–217

Dangas, G. and Kuepper, F. (2002) 'Restenosis: repeat narrowing of a coronary artery: prevention and treatment'. *Circulation*, 105, pp. 2586–2587

Dosi, G. (1982) 'Technological paradigms and technological trajectories: a suggested interpretation of the determinants and directions of technical change'. *Research Policy*, Vol. 11, pp. 147–162

Eeckhurst, E., Kappenberger, L. and Goy, J. L. (1996) 'Stents for intracoronary placement: current stats and future directions'. *JACC*, Vol. 27, pp. 757–765

Farooq, V., Gogas, B. D. and Serruys, P. W. (2011) 'Restenosis: delineating the numerous causes of drug-eluting stent restenosis'. *Circ Cardiovasc Interv*, 4, pp. 195–205

Fischman, D. L., Leon, M. B., Baim, D. S., Schatz, R. A., Savage, M. P., Penn, I., Detre, K., Veltri, L., Ricci, D., Nobuyoshi, M., Cleman, M., Heuser, R., Almond, D., Teirstein, P. S., Fish, R. D., Colombo, A., Brinker, J., Moses, J., Shaknovich, A., Hirshfeld, J., Bailey, S., Ellis, S., Rake, R. and Goldberg, S. for the Stent Restenosis Study investigators (1994) 'A randomized comparison of coronary-stent placement and balloon angioplasty in the treatment of coronary artery disease'. *New England Journal of Medicine*, 331, pp. 496–501

Fuchs, V. R. and Sox Jr. H. C. (2001) 'Physicians' views of the relative importance of thirty medical innovations'. *Health Affairs*, Vol. 20, No 5

Gelijns, A. and Rosenberg, N. (1994) 'The dynamics of technological change in medicine'. *Health Affairs*, Summer, pp. 28–46

Gelijns, A. C., Rosenberg, N. and Moskowitz, A. J. (1998) 'Capturing the unexpected benefits of medical research'. *New England Journal of Medicine*, Vol. 339, pp. 693–698

Gelijns, A., Sivin, J. G. and Nelson, R. R. (2001) 'Uncertainty and technological change in medicine'. *Journal of Health Politics, Policy and Law*, Vol. 26, pp. 913–924

Hamid, H. and Coltart, J. (2007) '"Miracle stents": a future without restenosis'. *Mcgill J Med*. 10(2), pp. 105–111

Kent, K. (2010) 'The history of interventional cardiology' in S. Redwood, N. Curzen, and M. Thomas (Eds) *Oxford Textbook of Interventional Cardiology*. New York: Oxford University Press, p. 22

King, S. B. III (1996) 'Angioplasty from bench to bedside to bench'. *Circulation*, Vol. 93, pp. 1621–1629

King, S.B. III (1998) 'The development of interventional cardiology'. *Journal of the American College of Cardiology*, Vol. 31, Supp. B, pp. 64B–88B

Li, J. and Kozlek (2012) 'The future of the drug eluting stent market: rebound will be driven by new, ground-breaking technologies and clinical confidence'. *Scientia Advisors Blog*. Available online at: www.scientiaadv.com/blog/2012/08/07/the-future-of-the-drug-eluting-stent-market

Mahmoudi, M., Delhaye, C. and Waksman, R. (2011) 'Safety and efficacy of drug-eluting stents and bare metal stents in acute coronary syndrome'. *Cardiovascular Revascularization Medicine* 12, pp. 385–390

Metcalfe, J. S., James, A. and Mina, A. (2005) 'Emergent innovation systems and the delivery of clinical services: the case of intra-ocular lenses'. *Research Policy* 34(9), pp. 1283–1304

Mina, A., Ramlogan, R., Tampubolon, G. and Metcalfe, J. S. (2007) 'Mapping evolutionary trajectories: applications to the growth and transformation of medical knowledge'. *Research Policy* 36(5), pp. 789–806

Mullins, C. B. and Lipscomb, K. (1977) 'Critique of coronary artery bypass surgery'. *Annual Review of Medicine*, Vol. 28, pp. 271–289

Mullins, S. M., Passamani, E. R. and Mock, M. D. (1984) 'Historical background of the National Heart, Lung and Blood Institute Registry for Percutaneous Transluminal Coronary Angioplasty'. *The American Journal of Cardiology*, Vol. 15, pp. 3C-6C

Myler, R. (2002) 'Coronary and peripheral angioplasty: historical perspective' Chapter 7 in Topol, E. (Ed.) *Textbook of Cardiovascular Medicine*. London: Lippincott Williams & Wilkins

Office of Technology Assessment (OTA) (1978) 'Assessing the efficiency of safety of medical technologies'. US Government Printing Office

Pepine, C. J., Hirshfeld, J. W., Macdonald, R. G., Henderson, M. A., Bass, T. A., Goldberg, S., Savage, M. P. Vetrovec, G., Cowley, M. and Taussig, A. S. (1990) 'A controlled trial of corticosteroids to prevent restenosis after coronary angioplasty'. *Circulation* 81, pp. 1753–1761

Ramlogan, R., Mina, A., Tampubolon, G. and Metcalfe, J. S. (2007) 'Networks of knowledge: the distributed nature of medical innovation'. *Scientometrics* 70(3), pp. 459–489

Reul, R. M. (2005) 'Will drug-eluting stents replace coronary artery bypass surgery?' *Tex Heart Inst J.* 32(3), pp. 323–330

Rosch, J., Keller, F. S. and Kaufman, J. A. (2003) 'The birth, early years and future of interventional radiology'. *Journal of Vascular and Interventional Radiology*, Vol. 14, pp. 841–853

Ruygrok, P. N. and Serruys, P.W. (1996) 'Intracoronary stenting: from concept to custom'. *Circulation*, 94(5), pp. 882–90

Serruys, P. W., de Jaegere, P., Kiemeneij, F., Macaya, C., Rutsch, W., Heyndrickx, G., Emanuelsson, H., Marco, J., Legrand, V., Materne, P., Belardi, J., Sigwart, U., Colombo, A., Goy, J. J., van den Heuvel, P., Delcan, J. and Morel, M. A. for the Benestent group (1994) 'A comparison of balloon-expandable-stent implantation with balloon angioplasty in patients with coronary artery disease'. *New England Journal of Medicine*, 331, pp. 489–95

Serruys, P. W., Kutryk, M. J. and Ong, A. T. (2006) 'Coronary-artery stents'. *New England Journal of Medicine*, 354, pp. 483–495

Simpson, J. B., Bains, D. S., Roberts, E. W. and Harrison, D. C. (1982) 'A new catheter system for coronary angioplasty'. *American Journal of Cardiology*, Vol. 49, pp. 1216–1222

Towers, M. and Davies, S. (2000) 'Cardiac catheterisation' in M. E. Silverman, P. R. Flemming and A. Hollman (Eds) *British Cardiology in the 20th Century*. London: Springer-Verlag, p. 151

Utterback, J. M. (1994) *Mastering the Dynamics of Innovation*. Boston, MA: Harvard Business School Press

Whitlow, P. L. (1997) 'Is rotational atherectomy here to stay?' *Heart* (Supplement 2), 78, pp. 35–36

Wilson, W. M. and Cruden, N. L. M. (2013) 'Advances in coronary stent technology: current expectations and new developments'. *Research Reports in Clinical Cardiology*, 4, pp. 85–96

World Health Organization (WHO) (2011) *Global Status Report on Non-communicable Diseases 2010*. Geneva: World Health Organization

Xu, S., Avorn, J. and Kesselheim, A. S. (2012) 'Origins of medical innovation: the case of coronary artery stents'. *Circulation Cardiovascular Quality and Outcomes*, 5: pp. 743–749

3 The evolution of the Left Ventricular Assist Device as a treatment for heart failure

Piera Morlacchi and Richard R. Nelson

Introduction

This chapter is about the origin, development and growing use of a device used to help heart failure patients, the Left Ventricular Assist Device, or LVAD for short. These devices are implanted in the body to assist failing hearts to do their pumping. Early versions of the LVAD began to be used in the late 1960s and since that time approximately 20,000 LVADS have been implanted around the world. As one would expect, today's LVADs are very different from the early versions. They are much less prone to malfunction, more effective in enabling patients to live longer, and easier for them to live with in reasonable comfort. The implantation procedure has become simpler, more routine, and less dangerous for the patient.

The initial chapter of this book discussed advances in medical practice more broadly and proposed that there were three different pathways associated with progress: the advance of scientific understanding of the human body and disease, the development of new technologies that could be used directly or indirectly to advance practice, and learning by experience in practice. The importance of these different routes to progress obviously differs depending on the type of advance. A large percentage of the cases analysed in this book involve the development of new devices that often enabled (and required) new techniques to be used by physicians. The LVAD is of this genre. (The book also includes studies of the development and employment of the artificial lens, angioplasty, artificial hips and discs.) In the case of LVAD, and also the other new devices considered in this book, the prior emergence of new technological capabilities that made the development of the device feasible was of critical importance. So too was learning in actual practice about what worked well and what didn't, with resulting changes in the procedures used, and also feedback to the engineers involved in design of the device. In the LVAD case advance in basic biomedical understanding was more something that came from experience with LVADS than something that motivated and oriented the development of the device.

The historical case of the evolution of the LVAD from its origins in the 1960s to today that is presented in this chapter is mainly based on archival materials and sources. The structure of the rest of this chapter is as follows. Sections 2 and 3 are the basic history of the evolution of LVAD as treatment for heart failure, with 2 giving a brief overview and 3 filling in the details. In section 4 we summarise what can be learned from the case of LVAD about advances in medical practice more broadly.

Contextual background

Advanced heart failure is a clinical syndrome that presents a collection of symptoms – fatigue, shortness of breadth, and congestion – that hits a person when her heart can not pump enough blood to satisfy her body's needs (Jessup 2001). In industrialised countries the improved management of coronary heart diseases short of heart failure and the improved longevity of the population more generally resulted in a growing number of patients with heart failure[1]. The evolution of treatment options for advanced heart failure patients over the last few decades has been impressive and now includes medical treatment based on drugs, heart transplantation and, most recently, mechanical circulatory device support therapies. The therapeutic innovation discussed in this paper is a device-based treatment for patients with advanced heart failure who are end-stage, i.e. they do not respond anymore to medical treatment and their only options are either heart transplantation or a mechanical circulatory device like the total artificial heart or the left ventricle assist device.

The treatment for failing hearts or heart failure in the 1950s would be considered primitive by today's medical standards. It was based on a combination of general measures - including dieting, limiting exercise and resting in bed for days or weeks - and a mix of drugs such as digitalis and injected mercurial diuretics, both known for their substantial toxicity. It was only at the end of the decade that new diuretics with more limited side effects became available. These diuretics are one of the many advances in treatment and technology for heart failure and other cardiac diseases that have been introduced since the 1950s (Silverman 1999). The 1950s had also seen development of the heart-lung machine, which made possible a new path of open-heart surgery, including not only repairing vessels, valves, and hearts but also replacing them. In 1953 the first successful open-heart surgical procedure demonstrated that the function of the heart and lungs could be replaced temporarily by mechanical means (Le Fanu 1999). Individuals have long dreamed about the possibility of replacing hearts and other organs of the human body that were beyond repair. Two alternative conceptions of what could be used to replace the failing heart of the patient have co-evolved. The first was heart transplantation, replacing the failing heart with a healthy natural organ provided by a deceased donor. The second was replacing it with a man-made or artificial heart. The natural heart of a donor was viewed as the best alternative, but both solutions were pursued in parallel.

In the early 1960s the dream of developing an artificial heart became a national priority in the US. Lobbied by the medical establishment the US Congress created the Artificial Heart Program in 1964. Modelled on the programme to land a man on the moon, it was based on the assumption that the artificial heart device could be brought into existence by using existing component technologies and that it would be developed in a few years (Hogness and Van Antwerp 1991).

> At the time, everyone believed that there was nothing we could not achieve utilizing modern scientific technologies. This was strongly influenced by the enthusiasm and the dream projected by former President John F. Kennedy. We were convinced that it was possible to develop a totally implantable cardiac prosthesis in the next 20 years.[2]

The initial plan presented by National Institute of Health (here after NIH) to Congress in their 1966 budgetary request called for both the development of cardiac replacement pumps or artificial hearts, and also for the development of other circulatory support devices to be used in emergency or acute situations as well as supporting failing hearts until transplantation.

The development of these mechanical circulatory devices occurred in parallel with the initial era of the heart transplants, which began with the first successful heart transplant procedure performed in December 1967. However, the hopes that heart transplantation could be the ultimate solution for patients with failing hearts were shattered quite quickly by the limited supply of donors, immunological problems that caused rejection of the received organs and severe complications in the few patients who received the heart transplants. Some of these problems were reduced by but not solved by the introduction of immunosuppressives in the 1980s, which reduced the problem of organ rejection (Frazier and Kirklin 2006), and by the creation of heart transplant lists and organ donors' networks that facilitate the allocation of hearts among transplant centers. Nonetheless, the problems encountered with developing heart transplantation to the point that it could be used more commonly in clinical practice, especially the limited number of donor hearts and the need to keep patients with failing hearts alive until transplantation, reinforced the demand and the expectations for the artificial heart (Fox and Swazey 1992).

However, the high hopes for the artificial heart were soon dampened. Initial tinkering, using existing technologies, revealed the complexity of this undertaking.

> 'We thought the problem wasn't going to be complicated as it turned out to be' he [William Pierce] says candidly, referring to the 'problem' of devising a safe and effective permanent artificial heart to replace ailing natural hearts. As we've solved some problems, we've uncovered others. That's why it remained a wonderful challenge[3].

It became evident that two big 'technical' problems existed: the lack of materials that could be accepted by the body and an appropriate source of energy. Another major limitation of the use of existing technologies to develop an artificial heart was the size of the overall device and its components, which constrained the possibility to totally implant the device in patients' bodies. Therefore, patients were bedridden and confined to live with those devices in hospitals. Moreover, the first patient who received an artificial heart, and the few others who followed, died of complications. Continuous media scrutiny exposed their misery. This early experience of using the artificial heart in human volunteers turned the public and some key opinion leaders in the medical community against the full-blown artificial heart technology as a solution to failing hearts. In the mid-1980s the hopes for rapid fulfilment of the dreams of greatly expanded use of heart transplantation and the development of an effective total artificial heart lay in tatters. The involved medical community largely refocused their attention and expectations instead on LVADs. The development of an LVAD increasingly came to be seen as much less demanding technically and thus a more quickly achievable solution for the problem of replacing and supporting failing hearts.

The LVAD is now viewed as a treatment option for patients with advanced heart failure who are not responding to medical therapy. This humble technology has come a long way and, in the following section, we consider the more detailed history of the evolution of the LVAD, fleshing out its developments after 1964, when the Artificial Heart Program was started by NIH in the US. Our history will be mostly about the US, although we will also recount events in Europe and in other parts of the world that are relevant to understanding US developments.

A detailed history of the origin and evolution of the Left Ventricular Assist Device (LVAD)

In this section we examine more deeply the history of LVAD that we sketched out in section 2. We cover the evolution of the device itself, advances over time of the clinical use of the LVAD as treatment for heart failure and, finally, the gains in scientific understanding of what is involved in heart failure and prognosis for recovery that have accompanied the advances of the device and the associated therapy.

The evolution of the device

Over the years LVAD devices have become more durable and reliable, smaller, simpler, easier to implant and more comfortable. The major developments that have enabled these improvements include: the development of materials compatible with the human body; technologies for components such as energy sources, storages, transmissions and converters; new instrumentation and improved methods to design and to test new devices. These developments were not independent and occurred over a number of decades. Many were made possible by exogenous advances drawn from broad technological areas, particularly electronics, computing and biomaterials, advances that enabled the design of improved component technologies for LVAD devices. In addition, there were, over the same period of time, improvements in surgical procedures and techniques to implant devices within the human body. In this sub-section we focus on the evolution of the LVAD devices.

LVAD as a device therapy for heart failure

To understand the origins and evolution of the LVAD device as treatment for heart failure it should be recognised that in the 1950s and 1960s the human body was viewed as a machine that had parts that could be repaired and replaced. The heart was seen as a pulsatile pump supporting the circulation of the blood within the human body and heart disease was understood to be caused by a diminished ability of the heart to pump blood. Broken or failing hearts were repaired by plumbing the heart through open heart surgery, which was made possible, as we have seen in the previous section, by the development of the heart-lung machine. It was believed that the only solutions for failing hearts that could not be mended through surgery and could not be replaced through heart transplants were mechanical blood pumps that could assist or replace these sick hearts. People believed that, because they knew how to build pumps, the goal of building mechanical blood pumps could be easily achieved in a short period of time and the problem of assisting or replacing broken or failing hearts could be readily solved. The common view was that two types of mechanical blood pumps could be developed as spare parts for broken or failing hearts: a 'total artificial heart (TAH)', a mechanical replacement of the whole natural heart that would be removed; and a 'ventricular assist device (VAD)', which was a mechanical circulatory support device that would assist the heart in doing its work (i.e. the natural heart itself would remain in place). The latter was in many cases a left ventricle assist device (LVAD) because the majority of the pumping work of the heart is done by the left ventricle.

The conceptualisation of the LVAD as a mechanical blood pump shaped the initial designs of the LVAD artifacts. Two families of designs started to appear from an early stage. First, temporary LVADs, i.e. blood pumps that could be used for a few hours or up to several days to bypass or augment the flow of the left ventricle in patients after cardiac damage, where all the components of the LVAD system, including the pump in some cases, were to be located outside the body, making the installation procedure much easier and safer, and enabling the pump to be easily removed when it was no longer needed (DeBakey 1971). Second, permanent LVADs, involving blood pumps where as many components as possible of the LVAD system (pump, energy converter, back-up battery, energy transmission receiver) were to be surgically inserted within the patient's body in order to take over most, if not all, of the work of the left ventricle for longer periods than several weeks (Kantrowitz 1990). A device of the first family of temporary blood pumps, the intraaortic balloon pump (IABP), which was developed in the 1960s, quickly gained broad clinical acceptance and it is still used in clinical practice as temporary cardiac support. We focus in this chapter on the second family of devices, permanent LVADs, most of whose components were designed to be implanted and to remain in the body for an indefinite time in support of failing hearts.

*Pioneering cardiac surgeons and engineers started to build mechanical
replicas of the heart (1950s–1960s)*

In the 1950s and early 1960s, people exploring the feasibility of an effective
LVAD were mostly cardiac surgeons and engineers. A number of research
groups led by pioneering cardiac surgeons started designing, experimenting on
animals and testing in a limited number of patients permanent and implanted
LVAD devices that were very primitive. They were prototypes, patched together
and fabricated with existing components by surgeons and their technicians, and
trialled in clinical practice through implantations in patients who were desperate
cases. Different device configurations were explored and redesigned based on
learning accumulated through animal studies and these initial implantations.
Moreover, these devices were a mechanical copy of the human heart (at least in
the part of the device that was implanted within the body). They were pulsatile
pumps, implanted in the body and connected directly to the heart in order to
enable blood flow to bypass the left ventricle[4]. The main components of the
device were the pump itself, the energy source and the control or drive unit, plus
other additional components depending on the specific design of each device[5].

These research groups were based mainly in medical centres with open-heart
surgery programmes. The developmental activities were conducted within experimental surgery and engineering laboratories, where electrical engineers were
employed to solve the problems identified by surgeons in the process of translating different ideas of LVAD designs into physical artifacts that were to be
implanted in the body (Bronzino 2005). In some cases, where this engineering
expertise was not available in-house or the problems encountered were beyond it,
the help of outside specialised engineers based in commercial companies was
sought to solve the problems encountered in the fabrication and redesign of these
devices. These *ad-hoc* collaborations resulted in the development of close and
personal relationships between a number of surgeons and engineers, based both
inside and outside medical centres (Kantrowitz 1990).

Moreover, significant cooperative work took place between different heart
centres. Many surgeons had known each other since their medical school training
or research fellowships under the mentorship of leading pioneers in the new and
emerging field of open heart surgery. Informal and cooperative relationships were
also created by frequent and face-to-face interactions. It was a common practice
to meet and argue about new ideas and practices at medical conferences and in
committees of professional associations and federal agencies, and for surgeons
and researchers to visit each other's laboratories when word of mouth spread
about new surgical techniques and/or new devices developed by an investigator.
Publications in the literature were used to share ideas but also to gain peer recognition and visibility in the medical field. Although competitive behaviours and
the desire to be the first in successfully attempting new surgical innovations and
make a name in the community was typical among cardiac surgeons, it was
considered in the interest of the innovator to allow other physicians to use and
evaluate the new technique independently in their own patients in order to gain

clinical acceptance of the new practice. One way to do this was to organise cooperative studies across a number of centres and funded through NIH grants (Kantrowitz 1990).

A limited number of research groups quickly became the hubs of this informal network among leading cardiac centres. In the early 1960s new artificial heart research programmes were formed at Pennsylvania State University, University of Utah, Baylor College and the Cleveland Clinic. In these institutions research on artificial heart technologies was conducted by large multidisciplinary groups of surgeons, cardiologists, physiologists, engineers, veterinarians and other medical and technical individuals. Their work included the design of devices, fabrication of prototypes and small batches of devices, bench evaluation, animal evaluation and clinical use. External collaborations with other academic institutions and commercial companies were established on specific investigator-initiated projects, funded in many cases by the NIH. Other medical centres such as the Children's Hospital in Boston, Stanford, and the Massachusetts General Hospital were engaged in validating and testing the functionality of the devices, and were developing techniques for implanting them, providing also feedback on possible adaptations of the artifacts.

Due to problems encountered in clinical applications such as the inadequacy of the materials to support patients beyond several days or weeks, the need for a major surgical procedure and the almost prohibitive cost of using the device clinically, early pulsatile devices were considered only as investigative procedures (DeBakey 1971). However, the possibility to further develop these early prototypes with the ambitious goal of using them in the clinical treatment of heart failure patients was becoming clear to an increasing number of people involved in the LVAD field.

More resources are needed: the creation of the Artificial Heart Program (mid 1960s)

The surgical investigators who were early entrants to the field played a pivotal role as the main advocates for securing federal support for their research and lobbied for the establishment of the Artificial Heart Program, which the US Congress established in 1964 and started funding in 1966. The National Heart, Lung and Blood Institute (NHLBI), through its Device Branch, was the NIH institute in charge of the Artificial Heart Program, with the task of directing and guiding the targeted effort to develop a family of mechanical circulatory devices for temporary and permanent applications to solve the national cardiac problems. In contrast with the majority of research funded by NIH, the mission-oriented and targeted research activities supported under the Artificial Heart Program, both on the total artificial heart and the left ventricular assist device, were financed through contracts awarded from targeted requests for proposal and not through grants to fund investigator-initiated projects (Altieri *et al.* 1986).

The continuity of funding in the late 1960s, based on contracts for the development of artificial heart technologies, stimulated the formation of new industrial research and development groups working in this area both within existing commercial companies and in newly established laboratories. For instance, companies working in the defense, space and instrumentation sectors like Aerojet and Thermo Electron Corporation started internal 'medical R&D groups' that were working mainly on NIH contracts. At the same time, industrial scientists and engineers started new small ventures like Avco Everett Research Laboratories, Andros Laboratories and Thoratec Laboratories to work on research and development of the artificial heart devices. Both the work in commercial companies and in these new laboratories was conducted in collaboration with surgeons and engineers based in medical centres and funded by federal contracts.

An annual conference was started at the beginning of the Artificial Heart Program by the NHLBI Device Branch for all investigators and contractors, based in medical centers, commercial companies and laboratories, which were funded by its contracts. The event was considered an effective and essential way to facilitate the sharing and dissemination of knowledge among the medical and engineering people working on developing a device treatment for heart failure.

A system approach to search for new component technologies (1970s)

By the early 1970s the hopes that a full blown artificial heart would be quickly developed were seriously compromised by the lack of progress on the problems of biocompatibility of materials and availability of appropriate energy sources encountered in the first decade of the Artificial Heart Program. It was believed that these problems would be less severe for the LVAD, so the NHLBI Device Branch decided to concentrate its limited resources on the development of this family of devices. This new direction in the development of a LVAD device came out of a broader consultation in the community involved, which NHLBI finalised in 1972. The Device Branch sponsored a conference to which many people involved in the field of mechanical circulatory devices were invited. The discussions during the conference emphasised the need for the development of long term and permanent LVADs and a shared view was that 'technology development for implantable assist devices would have application to all other forms of circulatory support' including the artificial heart (Altieri *et al.* 1986, p. 108). As a direct outcome of the conference a separate LVAD programme was created within NHLBI in 1974. The targeted effort of NHLBI was not only directed to the development of component technologies but also to the improvement of instruments and methods used in the off-line testing of new devices. Testing of a device starts with *in vitro* or bench test, followed by animal studies and then by use in humans. These systems are *in vitro* tests that simulate the body's environment, allowing the reliability and durability of devices to be evaluated[6]. For instance, the introduction of 'mock or artificial circulatory systems' in the early 1970s was a fundamental step in the development of the LVAD device due to its role in improving bench testing of devices.

The overall research and development effort of the field steered by the NHLBI Device Branch resulted in a number of LVAD devices designed and tested in the 1970s. These devices were the result of new collaborations between research groups based in commercial laboratories or companies and in academic medical centres that responded to NHLBI targeted requests for proposals and were awarded contracts to develop further the existing projects in their pipelines and to undertake new ones within the remit of the requests of proposals. In all of these pulsatile LVAD devices the blood pumps were made of polyurethane, implanted in the abdominal cavity (so the devices were called Abdominal LVADs or ALVADs) and connected to the heart by surgically opening the chest, with the connected console unit and energy sources left outside the body, with patients tethered to them and confined in beds.

The first group of devices to be tested in the 1970s were the ones originally developed in the late 1960s using existing component technologies. First results of the *in vitro* testing of the abdominal LVADs were reported in the literature, followed immediately by the results of tests in animals in early 1970s. The testing work in the lab with *in vitro* and *in vivo* on animals was fundamental for surgical innovators in order to develop experimental implantation techniques. Initial clinical trials in humans to evaluate the devices as a bridge to transplant or a device used to keep a patient alive in the period before a heart for transplant became available started in 1976.[7] The NHLBI sponsored a multi-centre trial at the Texas Heart Institute and the Children's Hospital Medical Center in Boston to evaluate the devices. These LVADs, built with old component technologies, were not very durable and reliable. However, the multi-centre trial and the cumulated clinical experience with these devices in other hospitals from 1960s through the early 1970s confirmed the clinical usefulness of LVADs in supporting the circulation of patients for longer periods than was possible with temporary blood pumps that could be used for only a few hours or up to several days.

Better implantation techniques, along with the need for better energy systems and biomaterials, were the main research challenges of LVAD programmes in the 1970s for the development of durable and reliable devices (Norman 1974). NHLBI supported the research activities of surgical innovators engaged in improving the implantation techniques of LVAD devices. These innovators were experienced and talented cardiac surgeons mainly located in medical centres with high volumes of heart procedures. An example of one of these medical centres is the Texas Heart Institute (THI), established in the early 1970s, which quickly became a central hub in the development and testing of new LVAD devices due to its cutting edge research on surgical procedures and the high volume of patients treated. Cardiac surgeons in medical centres like the THI were focusing their research activities on improving surgical techniques for the implantation of the devices, but also on testing and evaluating their performances in animals and in patients, providing valuable feedback - that, in some cases, only pointed out problems encountered in implanting the LVADs but, in other cases, also provided ideas on how to fix them - to commercial companies and laboratories that were designing and manufacturing these devices.

The LVAD becomes a temporary device for patients waiting for a heart transplant (1980s)

In the 1980s, the NHLBI Device Branch continued to play an active role in guiding the efforts of the LVAD community to solve existing problems of durability and reliability of the devices, which centred on energy sources and materials technologies. Different technological solutions were systematically explored and the most promising ones were selected out and incorporated as constraints in subsequent requests for proposals. An example of this approach is provided by the choice of electrical energy as the standard power source for LVAD devices in the 1980s. In this case, the Device Branch initiated a programme to develop implantable integrated electrically-actuated LVADs with two-year reliability. Contracts were awarded to the proposals for the development of pulsatile electrically-actuated devices or so-called 'first generation devices' presented by four commercial research groups[8]. In order to solve existing technological problems such as durability, reliability, complications caused by biomaterials such as bleeding and clotting, and improving the location for implanting the device, each group pursued different pulsatile designs of electrically-actuated LVADs. All these pulsatile blood pumps were big and heavy devices, requiring a lot of space to be implanted in the body, with two of the different devices still implanted in the abdomen and the other two implanted in the chest[9]. In addition to the devices developed under the four contracts awarded by NHLBI, other devices were in the pipeline of other commercial groups.

The design of these electrically-actuated pulsatile devices was conceived in mid 1980s with technologies and ideas inspired by the abdominal devices of the 1970s. Between 1975 and 1985 the devices 'underwent considerable development work to arrive to the design we are currently using' (Poirier 1997). The development and testing work by engineers over the bench and in the lab with animals performed well in *in-vitro* evaluation with mock circulatory systems, but it encountered unexpected problems when the devices started to be used in patients during clinical studies, which were initiated in mid 1980s. For example, one of these devices was highly reliable but introduced complications such as clotting that were not foreseen during its development and required further modifications. Other devices lacked reliability or were difficult to implant and required continuous modifications in response to feedback from the surgeons who were putting these pumps into patients. New protocols were designed to manage the complications of each pump in addition to fixing some of the technological problems encountered in using these devices in clinical trials. The problems were only partially due to technological choices embedded in the artifact. Additional factors that played a role in the outcomes of LVAD procedures included the reaction of each patient's body to the procedure, which in turn depended on a number of personal circumstances, and the quality of the implantation procedure, which surgeons were still in the process of developing and standardising for each device.

Notwithstanding the complexity involved in LVAD procedures, a number of first generation pulsatile electrically-actuated devices were successfully implanted in patients who were waiting for a heart transplant during the mid-1980s. To evaluate the progress of these pulsatile electrically-actuated devices, the NHLBI decided to sponsor an audited testing programme of the *in-vitro* readiness of these devices between 1986 and 1988. Altogether, the durability and reliability of four devices were evaluated. Because the LVAD devices were intended as a bridge to transplant, they needed to reliably support the heart for longer periods than a few days or weeks, so the Device Branch determined a two-year testing period. When one device successfully completed the evaluation, the NHLBI decided to sponsor the clinical trial of this device as a bridge to transplant, which started in 1991.

Pulsatile LVADs were the Model T of the industry (1990s)

All LVAD manufacturers used the learning generated by the two-years' *in vitro* test funded by NIH to modify their devices and to improve them in terms of durability and reliability through continuing clinical studies, sponsored by NIH or by a partnership between manufacturers and hospitals. In 1994 the FDA approved the first LVAD device as a bridge to transplant device[10]. Although this was an important moment in the evolution of the device, there was a high level of consensus in the field that major problems in the existing pumps remained. The devices were, on average, more durable and reliable, and their complications could be managed with different strategies such as drugs for blood clot prevention and the use of better biomaterials for the coating on the wires that penetrated the patient's skin, but they were still bulky and patients remained hospitalised.

In the mid 1990s, a lot of development work went into reducing the size of the LVAD devices. Electrically-actuated pulsatile devices were smaller than earlier abdominal LVADs. It was only with the introduction of rechargeable electrical batteries as the primary energy source, located outside the patient's body[11], that the average dimension of the LVAD device became comparable to a large format paper-bound book. The devices became 'implantable', i.e. the devices changed from being mostly outside the body, to being mostly implanted, leaving outside only the controller and the batteries, and 'portable' or un-tethered, i.e. the patients could leave the hospital to wait for transplantation. In 1997 the first LVAD patient could be discharged from the hospital and wait at home for their heart transplant. Overall, the lives of patients with end-stage heart failure were extended and made relatively better by LVAD devices, although not yet very comfortable due to the invasive surgical procedure necessary to implant the device and, after the implantation, the need to follow a drug therapy to manage complications, as well as the noise produced by the device and the need to regularly charge the electrical batteries. In 1998, the FDA approved two pulsatile LVAD devices to be used as bridge to transplant[12], which were fully implantable, electrically-powered, wearable LVAD devices. The regulatory approval for their use in clinical practice and commercialisation was given based on the provision of evidence of safety and effectiveness by the two manufacturers of the devices. This evidence was produced through multi-centre clinical trials conducted both in Europe and in the US in large-volume cardiac institutions.

For many people in the LVAD field, the two pulsatile devices were the Model T of the industry. They widely believed that more effective devices were getting ready to move from the research and development pipelines of academic laboratories and commercial companies to clinical trials in medical centres (Goldstein and Oz 2000). Moreover, some people came to the somewhat new view that LVAD could be a bridging technology but also a permanent treatment for end-stage heart failure. Surprisingly some of these people were cardiologists involved in the diagnosis and treatment of heart failure patients for whom the approval of the first fully implantable and wearable devices caused a dramatic shift in views about this therapeutic innovation. Since the beginning, for many cardiologists, even the ones involved directly with heart transplant centres, mechanical devices – both the artificial heart and the LVAD - were not real therapeutic options for patients with late-stage heart failure due to the poor quality of life that these pumps were offering. With the limited number of heart transplants due to the shortage of organs available, the only medical therapy they would consider for their heart failure patients was the one based on drugs. However, in the late 1990s, with the availability of new approved pumps some of them changed their view and started to consider LVAD as a therapeutic option.

In order to overcome problems of the first generation pulsatile devices, including the size of the devices and implantation complications, alternative technological trajectories for the development of new pumps started to be explored from the 1980s onwards. In the early 1990s, new innovative designs for non-pulsatile LVAD were developed and tested by different groups involved in the LVAD field. The development of these non-pulsatile designs, but also the continued work on pulsatile LVADs, was once again steered in part by the NHLBI. In 1994 the Device Branch issued a request for proposals for the development of 'Innovative Ventricular Assist Systems (IVAS)' to encourage innovation in the development of totally implantable ventricular assist devices. The NHLBI decided to fund proposals based on several different technological paths. Additional development for some of these devices was also supported from the mid-1990s by the NIH - Small Business Innovation Research (SBIR), a new federal funding scheme addressed in particular to small firms. An increasing level of commercial and financial interest in the LVAD market created a new wave of start-up companies, which was partially funded by the SBIR programme. Now that the proof of concept of the LVAD technology had been obtained and the LVAD therapy was increasingly accepted for patients with heart failure, much of the development work for the design of new LVAD devices was taking place in commercial companies. The IVAS programme, and other schemes like the SBIR and the entry of new commercial companies in the LVAD field, contributed to bring the new LVAD devices to a good stage of development and ready to be introduced in clinical studies. A very limited number of new non-pulsatile pumps started to be evaluated in medical centres in Europe and in Australia, where many LVAD manufacturers began to conduct their initial clinical trials in the early 1990s. In 1998 the first clinical application of continuous-flow assist devices as a bridge to transplant was done in the US. Clinical trials for many of the rotary LVADs in the research and development pipelines of different manufacturers really started in the US only in the early 2000s.

*'The LVAD devices are quite effective, but there are
too many of them!' (2000s)*

Over the last 10 years the views of some heart failure cardiologists, and of many other actors within and outside the LVAD community, regarding these devices started to change, since LVAD pumps have become more effective medical devices as they are, for example, reliable, smaller, simpler, more comfortable to live with and easier to implant. We now have a number of different LVAD devices with several still in development and in clinical trials, and with a minority approved for clinical use[13]. The smaller sizes of rotary pumps also opened up the possibility of developing devices for children, women and, in general, for people with a small body size in whom the larger pulsatile devices could not be implanted[14].

It is the opinion of some people within the LVAD community that there are presently too many devices with similar technical characteristics both in the market and in the pipelines of companies. This suggests a need to compare their real effectiveness through head-to-head clinical studies. Furthermore, the major problem that now needs to be tackled is not building another pump but reducing the complications or adverse events caused by LVAD implantation in the human body (Hunt 2007; Mussivant 2008), which cannot be eliminated only by developing more effective devices. This leads us to the discussion of the advances in the LVAD application procedures that have emerged and are currently employed.

Advances in the clinical procedures of LVAD

In the previous part of this section we described how LVAD moved from being an idea in the minds of a few surgeons and engineers to an effective device at the centre of a thriving community populated by medical centres, federal agencies, manufacturers, patients and other actors. We illustrated the interaction between the design and modification of the device and its use initially as an investigational or experimental procedure and then as an approved treatment for heart failure patients. We will now discuss the advances that occurred that made the LVAD an increasingly adopted life-saving therapy for the treatment of heart failure patients, starting with the improvements in the implantation procedure. At the present time this procedure remains a major cardiac surgical operation and this is likely to continue be the case for some time. It requires the skills of an experienced cardiac surgeon and a supporting surgical team, an appropriate infrastructure to conduct the operation and the pre- and post-operative care. Furthermore, it carries all the risks of complications and the recovery problems of a major cardiac surgical procedure[15].

The improvement of the LVAD implantation

When the LVAD pumps started to be implanted substantial experimentation was devoted to finding better ways to put them in human bodies and connecting them to the heart. In the 1960s and 1970s, the work to develop surgical innovations and new implantation techniques was left to experienced and talented surgeons who needed to figure out how to put pumps the size of a large format paperbound book or even bigger inside patients. Procedures were first tried out on animals in order to minimise the amount of experimentation on patients. Over time a number of different implantation techniques have been developed, refined and chosen over alternatives. For instance, the first generation pulsatile devices, which were large pumps, were placed in the abdomen of the patient. The new rotary devices now in clinical trials, due to their smaller sizes, are implanted near or within the heart (i.e. in the left ventricle) and they do not require the total opening the chest. The reduced size makes the implanting procedure easier and faster to perform, but also less invasive, so reducing the rate of surgical complications, making the recovery of the patient shorter and less painful. A further reduction in the size of the devices, which are now in development and offer partial support for the heart, allows for the possibility of the implementation of some of these devices through minimally invasive interventional techniques instead of major cardiac surgery.

Currently, the implantation techniques of different pulsatile and rotary LVAD devices involved in clinical trials are almost standardised procedures. Many cardiac surgeons involved in heart transplants are also the ones doing the LVAD implantations in both patients who ultimately receive transplants as well as other patients. A cardiac surgeon can perform an implantation with the skills acquired during surgical residency, complemented by further training with surgeons with LVAD experience and with specific knowledge of the device, which is normally supplied by the manufacturers. The basic steps required are the creation of a device pocket, the connection of the pump to the heart, the placement of the driveline and the actuation of the device. As in the case of other major surgical procedures, each surgeon has their own preferences for the ways in which to conduct some of these steps (for example, choosing to do the implantation with or without the use of cardiopulmonary bypass) (Morris 2007).

The development of pre- and post-operation procedures,
LVAD teams and centres

The competence and experience of the heart surgeons who implant LVADs is a necessary but not sufficient condition for successful outcomes from the procedure. A team of professionals needs to support the surgeon during the operation, but also in the pre-operative and post-operative care stages. This support includes explaining the risks and benefits involved to possible patients and their families of the LVAD as therapeutic option and managing the complications of the implantation and the management of the devices. The core LVAD group normally includes, besides the cardiac surgeon, a bioengineer in charge of monitoring the LVAD device, a coordinator to take care of protocols and hospital logistics and a physician who could follow up and manage the patient's daily conditions, including nutrition, exercise, medications and various tests (Oz 1998). In many centres this physician, the 'transplant or LVAD doctor' or the heart failure cardiologist, refers the patient to the surgeon for a LVAD procedure and follows it up afterwards. Additional members of these normally multidisciplinary LVAD teams include nurses, social workers, physical therapists, physician assistants and psychologists.

In early 1990s, with the first generation devices approved as a bridge to transplant for end-stage heart failure patients, a number of LVAD programmes were created in existing heart transplant centres with significant volumes of patients treated. The management of the patients undergoing LVAD implantation and heart transplant has some similarities, in addition to the fact that the same patients were implanted with an assisted device in order to be transplanted later on. Additional new programmes were set up subsequently by many hospitals with existing heart programmes[16]. Leading LVAD programmes manage a small subset of the devices currently available (both under experimentation/trials or for approved clinical use) because there is continuous learning derived from the use of each of these devices and the development of the LVAD therapy is still on ongoing process. Furthermore, new procedures are currently being developed in order to refer and screen patients who are good LVAD candidates, but also to design and to manage outpatient programmes, i.e. services for implanted patients who leave the hospital to return home. Nutrition, exercise and complementary drug therapy are examples of additional factors in the pre- and post-operative care stages that contribute to better outcomes in the overall LVAD therapy.

Improving outcomes, expanding indications and proving effectiveness

A significant step in establishing a therapy for heart failure patients based on the LVAD was in demonstrating its effectiveness in comparison with existing treatment modalities. The REMATCH trial (1998–2001)[17] involving first-generation LVADs implanted in end-stage heart failure patients was a major step forward in the investigation of LVAD as therapy for heart failure and it showed that early survival of patients on LVADs doubled when compared with those on medical therapy at one year. It generated solid and robust evidence about the safety and effectiveness of LVAD in very sick patients – the therapy supports functional survival of patients, establishing LVAD as 'destination' or permanent therapy for these patients (i.e. the last hope for dying patients with 'one foot in the grave').

Since the REMATCH trial, the indications for LVAD therapy have changed and expanded, although the progress of establishing LVAD as therapy for other than heart failure patients at extreme end stage proceeded at a slow pace. The availability of smaller devices less prone to technological failures, i.e. more reliable, and the improved experience with implanting procedures contributed to this progress. The key problem now is to characterise the target population that can benefit from LVAD beyond end-stage heart failure. The expansion of the inclusion criteria for mechanical support in recent years towards less sick patients has been important[18]. Several reports have sought to identify significant pre-operative variables that may predict risk and impact outcomes (for example, in terms of reducing of mortality and/or complications). For example, the timing of intervention has proven to be an important determinant of clinical outcomes. Risks and complications can also be viewed as the result of treating patients who are too sick to be treated: 'If we could operate on these patients earlier, our results would be better'. It is a vicious circle that is difficult to break: patients who are treated are too sick, so putting LVAD in them carries additional risks and complications compared with the ones associated with the device and the characteristics of patients (i.e. who receive the implantation after their nutritional status and end-organ function have declined too much). Cardiologists who are treating heart failure patients view the LVAD as a last resort and refer patients when they are too sick for other treatments.

The NHLBI Device Branch, in cooperation with the FDA and the Center for Medicare and Medicaid Services, is trying to monitor changing effectiveness and safety outcomes. The three federal institutions have supported the creation of a clinical data registry – called Intermacs (Interagency Registry of Mechanically Assisted Circulatory Support) – to cumulate outcomes and other data on LVAD patients across different centres and implanted with different devices. Initially, this register was built by professional medical societies, which were trying to encourage manufacturers to share data about devices and patients that were initially cumulated in proprietary registries only available to their investigators. The formation of a registry was not welcomed by manufacturers who viewed this as an effort to harm competition among different devices and companies. At the beginning participation in the Intermacs registry was left at the discretion of manufacturers and investigators, whereas it is now enforced through a cooperation in which FDA and CMS link the approval and reimbursement processes of new devices to the participation in the registry. This serves to stimulate learning at the level of the field across devices, in the selection and management of patients for example. The significant improvement in the quality of the clinical outcomes in the use of LVADs, as evidenced by the REMATCH study, certainly provides an argument for their more general use. However, the cost of treatment using LVAD continues to be very high, which certainly is and should be a deterrent to their diffusion. While the experience cumulated by some LVAD programmes with a significant volume of patients shows that further reductions in the cost of LVAD therapy are possible, this hasn't happened yet. At present, both because of its high cost and because many cardiologists are not yet persuaded of its benefits to heart failure patients, the LVAD therapy is still not widely adopted as treatment for eligible patients by the wider medical community involved in the treatment of heart failure. Although the LVAD was included as a therapy for end-stage heart failure patients in the medical guidelines published in 2005 by the American College of Cardiology (Hunt *et al.* 2005), the clinical implementation of guidelines in the LVAD case, as in other medical therapies, is still limited and eligible patients are not referred to this therapy, showing that a number of barriers still need to be overcome.

In the next sub-section we discuss how the use of LVAD treatment in clinical practice was instrumental in advancing the understanding of heart failure and the diseases that produce it. Perhaps this more recent development in the LVAD story could lead to a larger adoption of this therapy, based on the legitimacy provided by relating its advancement to scientific understanding of the heart failure diseases.

Advances in the scientific understanding of heart failure

> Despite repeated attempts to develop a unifying hypothesis that explains the
> clinical syndrome of heart failure, no single conceptual paradigm for heart
> failure has withstood the test of time.
>
> (Mann 1999.)

Over the last century we have developed a much better scientific understanding
of the heart, which, in some cases, has been instrumental in developing treatments for some of its diseases. For instance, an understanding of the anatomy and
physiology of the heart was important for the development of surgical treatments
for vascular diseases. Heart failure in clinical practice is diagnosed based on a
number of measures of the cardiac output or capacity of the heart to pump blood
and treated with a combination of approaches that includes drugs, surgery,
devices, nutrition, exercise and rest. But these treatments for heart failure, as in
the cases of other heart diseases, have been developed since the 1950s without a
deep scientific understanding of the causes and mechanisms of this disease.

A scientific model of the causes and mechanisms of heart failure largely
accepted by the medical community is the hemodynamic model, where the
disease is thought to arise as a result of abnormalities of the pumping capacity of
the heart or excessive peripheral vasoconstriction as in obstructed arteries.
However, we do not know what causes these abnormalities in the first instance.
Furthermore, this hemodynamic model does not, for example, explain the
progression of heart failure in patients treated with current medical therapy
(Mann and Bristow 2005). 'Heart failure has a worse prognosis than most cancer,
but heart failure lags far behind cancer in the robust staging of patient profiles and
prognosis' (Warner Stevenson and Couper 2007, p. 750), because we do not have
a detailed understanding of the factors behind the risk, mechanisms and progression of heart failure in patients. As in the case of other treatments for heart failure,
the LVAD therapy has been developed without a detailed scientific understanding
of the causes and mechanisms of this disease. On the contrary, a lot was learned
about the heart relevant to LVADs and beyond in the course of developing and
using LVADs. This is the case in the recovery of the heart: we now know that
heart failure can be reversed by providing prolonged LVAD support to the heart,
which can in this way temporarily download its pumping function and support its
recovery. Cardiac clinicians and researchers did not think that the heart could
recover from heart failure, although this idea was initially proposed in the 1960s
by Burch who stated that 'the heart, like any other diseased organ, improves with
rest' (1966, p. 422). The possibility of cardiac recovery during ventricular assist
device support was reported anecdotally in the literature in 1994 by Frazier and
was observed later on by other researchers.

> While end-stage heart failure was once thought to be irreversible, research
> now suggests that LVAD support may lead to both cellular and functional re-
> covery. Ultimately, patients with advanced cardiac disease might be managed

with temporary mechanical support combined with pharmacological and cellular therapies, in place of cardiac transplantation or long term LVAD support.
(Maybaum *et al.* 2008, p. 234.)

It has now been demonstrated with sufficient evidence that the heart can recover in some cases, and even completely when supported for a prolonged period of time by a LVAD. In the last ten years a major focus of investigation in the field of advanced heart failure research became the temporary use of LVADs as bridge to recovery of the patient's own heart. Both basic and clinical heart failure scientists are now working to understand and explain the cumulated clinical evidence and the underlying cellular and structural mechanisms for cardiac recovery seen in patients with prolonged LVAD support (Torre-Amione and Loebe 2006). Moreover, the combination of LVAD implantation with new pharmacological therapy and stem cells that can increase the chances of success in promoting the recovery of the heart in a larger number of patients is now being investigated (Birks *et al.* 2006). The contribution made by the LVAD treatment of heart failure in improving the scientific understanding of heart disease could lead to its widespread dissemination as adopted medical practice.

The evolution of the LVAD as a treatment for heart failure

Overall, for some people, the LVAD is now ready for the 'prime time' in the clinical treatment for heart failure after a journey of 50 years (McCarthy and Smith 2004): the device is quite effective, especially the new devices that are now in the pipeline and are reaching clinical trial stage; the LVAD practice is clinically used, although not widely diffused, to treat heart failure; and new scientific understanding about heart failure disease is coming out from the development of the device and its clinical use in patients. As LVADs have improved, they have increasingly been implanted as a treatment for heart failure in their own right, rather than as a bridge to transplant. Whereas before there was a considerable period when knowledgeable people could argue that an LVAD implantation did not generally improve the conditions of life of a patient, the most recent evaluations indicate that the current LVADs (or, more accurately, the earlier LVAD implantations that were evaluated) do on average extend the life of the patient and that living with an LVAD, while still not easy, is more comfortable than it used to be. There are strong indications that in the coming years LVADs will become more reliable, be much smaller, easier to implant and to live with, maybe even cheaper or cost effective, although we are more sceptical about the latter.

The LVAD case shows vividly the fabric and the dynamics of the three mutually constituting, interdependent and evolving pathways involved in the advance of medical therapy when a medical artifact is involved: enhanced ability to design and develop effective medical artifacts like new drugs and medical devices of various sorts; improvements in the medical procedures employed by physicians and healthcare organisations themselves; and advances in scientific understanding of disease and body function. The efforts that led to today's LVADs were not induced by any advance in understanding of heart disease. Rather, the efforts started in the 1960s because of the general view that the basic technology was then available for the development of an artificial heart, an effort that gradually shifted over to the development of heart assist devices. Nor, over the course of these efforts to date, has there been any major advance in basic biomedical understanding that has helped to guide the effort. On the other hand, experience with patients who have had LVADs implanted has led to an enhanced ability to design and develop new medical technologies, and to a new understanding that a damaged heart can at least partially heal itself if relieved of some of the pumping burden needed to sustain life. While not from basic scientific research, a lot has been learned over the history of LVAD that has helped the evolution of this new medical therapy.

Acknowledgement

This chapter contains excerpts reprinted from Morlacchi, P. and Nelson, R. R. (2011) 'How medical practice evolved: learning to treat failing hearts with an implantable device'. *Research Policy*, 40(4). Copyright 2011, with permission from Elsevier.

Notes

1 In the US and in Europe, with over 700 million inhabitants, the estimated incidence of heart failure is around 7 million, and the prevalence of advanced heart failure is estimated to total between 70,000 and 700,000 patients (Deng and Naka 2007).
2 Pierce quoted in Berkowitz (1990).
3 Pierce quoted in Berkowitz (1990).
4 The device was removing blood through the ventricular apex and returning it to the abdominal or thoracic aorta.
5 For instance, internal batteries, inflow and outflow grafts to connect the pump to the heart, air drive lines, etc.
6 The mock circulatory system that became *de facto* the NIH standard for LVAD testing was developed at Penn State in 1972 based on research funded by NHLBI.
7 It is important to notice that the Medical Device Amendment Act in 1976 changed the existing practices to develop and test LVAD devices. The use of implantable medical devices like LVADs in patients both for research and clinical use became regulated by the Food and Drug Agency (FDA). Whereas before the passage of this act the use of experimental devices in humans was monitored by local advisory committees, after 1976 these decisions were centralised and supervised by the regulatory federal agency.
8 These groups were: Andros/Novacor, Thermedics/TCI, Thoratec, Avco-Everett/Abiomed.
9 The Abiomed was the largest at 1210cc in implanted volume and 2.2 kg of component weight and the Thermedics the smallest respectively at 760cc and 1.3 kg.
10 The first device approved was the Heartmate by Thermedics.
11 The external batteries were connected via a lead going through the patient's skin to the inside pump.
12 The Thermedics device called HeartMate I and the Novacor device.
13 In 2007 there were seven or more LVADs in different stages of clinical trials in the US.
14 In 2004 NHLBI funded five contracts for five years to develop paediatric LVADs.

15 For instance, limited risks of death and of neurological damages when the cardiopulmonary bypass is used plus pain to the chest and time to recover after the surgery.
16 In some cases these centres do not have a transplant programme.
17 The clinical trial started in 1998 and results were published by the main investigators for the first time in 2001 in the highly prestigious *New England Journal of Medicine* (see Rose *et al.* 2001).
18 LVAD is now used for a larger population of patients with heart failure, generated by both acute and chronic cardiovascular disease. The devices can be used as a bridge to transplantation and as destination therapy in patients with end-stage heart failure, but also as a bridge to recovery as described in more detail in the next sub-section.

References

Altieri, F. D., Watson, J. T. and Taylor, K. D. (1986) 'Mechanical support for the failing heart'. *Journal of Biomaterials Applications*, 1, pp. 106–156

Berkowitz, S. I. (1990) 'The beat goes on' at the symposium on *The Artificial Heart: Past, Present and Future. Honoring William S. Pierce.* Hershey Medical Centre, Penn State University

Birks, E. J., Tansley, P. D., Hardy, J., George, R. S., Bowles, C. T., Burke, M., Banner, N. R., Khaghani, A. and Yacoub, M. H. (2006) 'Left Ventricular Assist Device and drug therapy for the reversal of heart failure'. *New England Journal of Medicine*, 355, 18, pp. 1873–84

Bronzino, J. (2005) 'Biomedical engineering: a historical perspective' in J. D. Enderle (Ed.) *Introduction to Biomedical Engineering.* Amsterdam: Elsevier

Burch, G. E. and DePasquale, N. P. (1966) 'On resting the human heart'. *Am Heart J*, 71(3), p. 422

DeBakey, M. E. (1971) 'Left ventricular bypass pump for cardiac assistance'. *The American Journal of Cardiology*, 27, pp. 3–11

Deng, M. C. and Naka, Y. (2007) *Mechanical Circulatory Support Therapy in Advanced Heart Failure.* London: Imperial College Press

Fox, R. C. and Swazey, J. P. (1992) *Spare Parts. Organ Replacement in American Society.* Oxford, UK: Oxford University Press

Frazier, O. H. (1994) 'First use of an unthetered, vented electric Left Ventricular Assist Device for long-term support'. *Circulation*, 89, pp. 2908–2914

Frazier, O. H. and Kirklin, J. K. (Eds) (2006) *Mechanical Circulatory Support.* Amsterdam: Elsevier

Goldstein, D. J. and Oz, M. C. (2000) *Cardiac Assist Devices.* Armonk, NY: Futura Publishing

Hogness, J. R. and Van Antwerp, M. (1991) *The Artificial Heart: Prototypes, Policies, and Patients.* National Academy of Sciences

Hunt, S. A., Abraham, W. T., Chin, M. H., Feldman, A. M., Francis, G. S., Ganiats, T. G., Jessup, M., Konstam, M. A., Mancini, D. M., Michl, K., Oates, J. A., Rahko, P. S., Silver, M. A., Stevenson, L. W., Yancy, C. W. and ACC/AHA (2005) 'American College of Cardiology/American Heart Association 2005 guideline update for the diagnosis and management of chronic heart failure in the adult'. *Circulation*, 112, pp. 154–235

Hunt, S. A. (2007) 'Mechanical circulatory support. New data, old problems'. *Circulation*, 116, pp. 461–462

Jessup, M. (2001) 'Mechanical cardiac-support devices: dreams and devilish details'. *New England Journal of Medicine*, 345, 20, pp. 1490–1493

Kantrowitz, A. (1990) 'Origins of intraaortic balloon pumping'. *Annals of Thoracic Surgery*, 50, pp. 672–674

Le Fanu, J. (1999) *The Rise and Fall of Modern Medicine*. London: Abacus

Mann, D. L. (1999) 'Mechanisms and models in heart failure. A combinatorial approach'. *Circulation*, 100, pp. 999–108

Mann, D. L. and Bristow, M. R. (2005) 'Mechanisms and models in heart failure: the biomechanical model and beyond'. *Circulation*, 111, pp. 2837–2849

Maybaum, S., Kamalakannan, G. and Murthy, S. (2008) 'Cardiac recovery during mechanical assist device support'. *Seminars in Thoracic Cardiovascular Surgery*. 20(3), pp. 234–46

McCarthy P. M. and Smith, W. A. (2004) 'Mechanical circulatory support. A long and winding road'. *Science*, 295, pp. 998–999

Morris, C. R. (2007) *The Surgeons. Life and Death in a Top Heart Center*. New York: Norton

Mussivant, T. (2008) 'Mechanical circulatory support devices: is it time to focus on the complications, instead of building another new pump?' *Artificial Organs*, 32, 1, pp. 1–4

Norman, J. C. (1974) 'An Abdominal Left Ventricular Assist Device (ALVAD): perspectives and prospects'. *Cardiovascular Diseases, Bulletin of the Texas Heart Institute*, 1, 3, pp. 251–264

Oz, M. C. (1998) *Healing from the Heart*. New York: Plume

Poirier, V. L. (1997) 'The Heartmate Left Ventricular Assist System: worldwide clinical results'. *European Journal of Cardio-Thoracic Surgery*, Sup. 11, pp. S39-S44

Rose, E. A., Gelijns, A. C., Moskowitz, A. J., Heitjan, D. F., Stevenson, L. W., Dembitsky, W., Long, J. W., Ascheim, D. D., Tierney, A. R., Levitan, R. G., Watson, J. T., Meier, P., Ronan, N. S., Shapiro, P. A., Lazar, R. M., Miller, L. W., Gupta, L., Frazier, O. H., Desvigne-Nickens, P., Oz, M. C. and Poirier, V. L. (2001) 'Long-term Mechanical Left Ventricular Assistance for end-stage heart failure'. *New England Journal of Medicine*, 345, 20, pp. 1435–43

Silverman, M. E. (1999) 'A view from the Millennium: the practice of cardiology circa 1950 and thereafter'. *Journal of American College of Cardiology* 33 (5), pp. 1141–1151

Torre-Amione, G. and Loebe, M. (2006) 'Myocardial recovery following prolonged mechanical support' in O. H. Frazier and J. K. Kirklin (Eds) *Mechanical Circulatory Support*. Amsterdam: Elsevier

Warner Stevenson, L. and Couper, G. (2007) 'On the fledging field of mechanical circulatory support'. *Journal of American College of Cardiology*, 50, 8, pp. 748–751

4 Uncertainty in the hybridisation of new medical devices

The artificial disc case

David Barberá-Tomás

Introduction

This case study of the intervertebral artificial disc affords the opportunity to analyse two kind of uncertainties related to new medical device designs. The uncertainties refer here not only to basic medical knowledge (as is usual in studies of innovation in medical devices, see Metcalfe *et al.* 2005, Mina *et al.* 2007, Morlacchi and Nelson 2011) but also to the performance of devices whose design is based on different fundamental understandings of disease. In this work, I argue that the hybridisation of devices is one of the strategies used to manage the latter uncertainty. Hybridisation consists of the embodiment of multiple and functionally competing operational principles[1] within a single medical device when there is little or no clinical data on their comparative performance. Stated in conditional logic language, the justification for hybridisation is that: 'if you do not know which operational principle is better, then choose all'.

The artificial intervertebral disc is implanted in the spine to treat Degenerative Disc Disease (DDD), a physiological affliction of the intervertebral discs.[2] It is recognised as the main cause of back pain and disability among adults in the United States (Errico 2005), and its associated costs are estimated at around $26 billion in the US (Duke University Medical Center 2004) and £1.6bn in the UK (Maniadakis and Gray 2000).

The artificial disc is an alternative to arthrodesis, or osseous fusion – the gold standard for the treatment of DDD, consisting of the replacement of the disc with an osseous bridge between adjacent lumbar and cervical vertebrae. Figure 4.1 depicts the basics of the fusion (arthrodesis) and artificial disc replacement (arthroplasty) therapies. The left side of the figure 'shows a series of vertebrae 50, 51, 52, 53, in which the vertebrae 51 and 52 have been fused together ... [and] the [natural] disc ... removed. As a result, the adjacent discs 54 and 55 have to flex considerably. The [right] side [of the figure] shows a series of vertebrae 60, 61, 62, 63, with the prosthetic disc 10 between the vertebrae 61 and 62. As a result the adjacent discs 64 and 65 have to flex considerably less to reproduce the

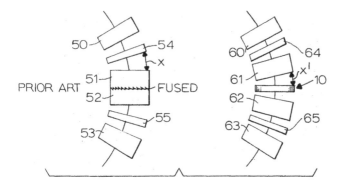

Figure 4.1 Left: the fused vertebrae in an arthrodesis. Right: the degenerated disc replaced
       by a prosthesis.

Source: US3867728.

equal amount of spinal flexion, the intervertebral distance X' in the invention
obviously being less than the corresponding distance X in the [fusion
procedure]'.

This fragment from the text of patent no. US3867728 (one of the first artificial
disc patents) explains the medical problem at hand. The goal of the fusion procedure is to extract the natural disc affected by DDD (the source of pain and disability) and substitute it with an osseous bridge. What the artificial disc adds to this
outcome is restoration of the natural motion in that vertebral level, which is seen
as an 'intuitively attractive alternative to spinal arthrodesis' (Freeman and
Davenport 2006, p. 440). We show that some specific designs of artificial discs
were conceived to add to this 'intuitive' advantage of arthroplasty over fusion by
the restoration of the load absorption capacity of the natural disc, something that
is far from being intuitive.

Later in the chapter (section 2 onwards) we discuss the hybrid design strategies
adopted in the face of scientific and clinical uncertainty surrounding how much
load an artificial disc would be required to bear. We would emphasise another
important element referred to in the extract from the patent reproduced above: the
deleterious systemic effect of osseous fusion in the functioning of the rest of the
spinal column. As the extract highlights, the deleterious effects of fusion are not
only absence of mobility in the corresponding vertebral level but also an abnormal amount of mobility in adjacent vertebral levels ('x', on left side of
Figure 4.1). Some claim that this abnormal motion ultimately causes degenerative
disease in the adjacent discs (the so-called 'adjacent disc degeneration syndrome')
and leads to the need for further surgery (Lee and Goel 2004). There is also an
analogous systemic effect (discussed below) that adds to the uncertainty
surrounding load absorption.

We provide a brief description of the history of the spinal implants used in fusion/
arthrodesis (screws and bars) and disc replacement/arthroplasty (artificial discs).

Arthrodesis in DDD treatment began as a non-instrumented surgical therapy, consisting of spinal fusion between vertebrae achieved without the help of any implantable devices (Figure 4.1). The use of screws and bars to achieve better fusion of adjacent vertebrae originated in Europe in the 1980s and gained momentum with the publication of results confirming better fusion (Boos and Webb 1997). These implants had been conceived in the early 1960s for the surgical treatment of other spinal pathologies such as fractures. From the early 1990s the use of pedicle screws for DDD surgical treatments was the main driver of spine surgery and, for the last two decades, this has been a fast-growing device sector in the orthopaedic industry, reaching more than $6 billion in 2007 (MDDI 2008) and overtaking traditional leaders as hip and knee prostheses.

Diffusion of the artificial disc has been slower. Although experimental artificial disc surgeries were being performed in Europe in the mid-1980s (almost contemporaneously with the first uses of spinal screws for DDD surgical treatment), its diffusion did not gain momentum until the early 2000s. Since then artificial discs have experienced considerable growth and are estimated to be worth $3 billion dollars. They have progressively overtaken arthrodesis as the standard clinical practice in the surgical treatment of DDD (Lieberman 2004; Singh *et al.* 2004). There are several possible explanations for the slower diffusion of the artificial disc as compared with spinal screws. First, proving the efficacy of the whole artificial disc rationale compared with the dominant technique of spinal fusion was difficult: in randomised studies, although disability and quality of life scores tipped in favour of artificial disc procedures, there was no conclusive evidence supporting either procedure. In its evidence-based guidance on the use of the artificial disc issued in 2009 and 2010, the UK National Institute for Health and Clinical Excellence (NICE) identified five randomised controlled trials comparing the efficacy of artificial disc and fusion. In four of these studies differences were found to be non-significant at follow-up. This affected the launch of the artificial disc in the US, where many insurance companies, as well as Medicare and Medicaid, initially authorised very limited or no reimbursement for the procedure in the United States 'due to the lack of good evidence of long-term clinical benefit and safety' (Kurtz 2006, p. 329). This was even after the FDA-approved trials. Although reimbursement for artificial discs has increased, it remains an important barrier to the widespread adoption of this technology in the US. Another reason is that the spinal screws technique was already in use for other spinal pathologies such as fractures. This facilitated their diffusion to DDD surgical therapy, whereas the artificial disc technique involved new and risky surgical approaches (Mayer 2005).

In addition to these clinical reasons, industrial development of the artificial disc experienced many false starts and hitches. A condensed history of the two most important early artificial disc development projects is illustrative. Attempts to use primitive artificial disc designs began in the mid-1960s by pioneer surgeons. However, it was not until the 1980s that industry began to be involved in the development of these prostheses. The SB Charité artificial disc was first experimentally implanted in East Germany during early 1980s with bad results that were attributed to the design and manufacturing techniques employed. In the mid-1980s Waldemar Link, a West German company, joined the project and in 1987 began commercialisation of a considerably improved design in France, Netherlands, and West Germany. This was the first time an artificial disc had been marketed (Geisler 2006). However, the early bad results from the first SB Charité prototypes had a long-lasting influence on many surgeons' perceptions of the artificial disc (McAfee 2003). In the 1990s, a sequence of events helped to partially reverse this situation and boosted artificial disc sales in early 2000 (Figure 4.2). In 1994 the first positive results from non-randomised trials using the improved SB Charité were published (Griffith *et al.* 1994). Four years later the Frenchay hospital in Bristol, UK, a prestigious centre in the field of neck surgery, announced positive experimental (also non-randomised) results on humans for a cervical disc based on the SB Charité operational principle (Cummins *et al.* 1998). In 2000, the FDA's announcement of approval for a randomised clinical trial of SB Charité was the first sign of the opening of the lucrative US market to artificial discs. The US market represents more than half of the world market. FDA final approval of SB Charité in 2004 was the first approval for a prosthesis of this kind for clinical use in the US.

The history of the other important artificial disc project developed during the 1980s did not have a positive outcome. The Acroflex artificial disc was developed by Acromed, an Ohio company founded by the spinal surgeon Arthur Steffe. The project started in the mid-1980s and involved at least three series of laboratory and experimental human testing, all of which produced bad results (Kurtz 2006). The project was abandoned in 2000. Section 2 discusses the scientific and clinical uncertainties involved in the SB Charité, Acroflex and other artificial disc development projects. Section 3 discusses design hybridisation as a response to these uncertainties.

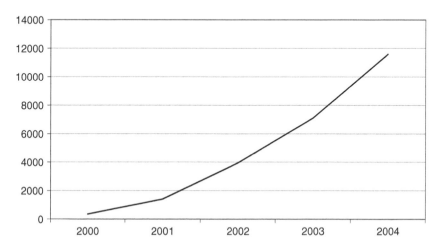

Figure 4.2 Evolution of sales (in US$) of artificial discs in 2000–2004.

Source: own elaboration of data in Biondo and Lown 2004.

Uncertainty and the artificial disc

The specialist scholarly literature offers two different views on the biomechanical nature and pathogenesis of DDD (Bono and Garfin 2004).[3] Kinematic theory focuses strongly on the movement of the spinal disc but does not take account of the forces that produce the motion. Dynamic theory is concerned mostly with the combined effect of motion and loads.[4] According to the first theory, back pain is a consequence of abnormal movements in the disc affected by DDD, and artificial disc replacement restores normal mobility (Mulholland 2008). According to dynamic theory the dynamic properties of the anatomic disc, and specifically the load absorption of the cartilaginous articulation, play a crucial role in triggering the disease. It claims that to mimic anatomic load absorption, the artificial disc should reproduce the viscoelastic properties of a healthy disc. Both the kinematic and dynamic theories about the sources of back pain consider intervertebral fusion or arthrodesis as an inappropriate surgical treatment for DDD, arguing that the rigid nature of the bone bridge could trigger biomechanical alterations that, ultimately, could cause degeneration in adjacent discs and result in the need for further surgery (Fekete and Porchet 2010; see Figure 4.1).

Although early hints about the dynamic aspects of disc functionality date back to the early 1970s (Urbaniak *et al.* 1973), uncertainty persists about the actual importance of load absorption. The difficulties involved in measuring load absorption both *in-vivo* and in laboratory environments, are the main cause of the paucity of data on the properties of the intervertebral disc (Le Huec *et al.* 2003, p. 347). To our knowledge, only one laboratory study has tried to analyse the impact of load absorption using invasive force sensors installed in a cadaveric model of spine units (Dahl *et al.* 2006)[5]. Deficient knowledge about the workings of the natural disc has had a crucial effect on the designs of an artificial device to replace it. The different R&D projects on artificial disc treatments tell the story of how this uncertainty became an essential element of the various designs solutions, according to different designers' notions about how much load an artificial disc is required to bear.

The first artificial disc to be adopted for clinical use was built on what we refer to as the 'hip-like' operational principle, a ball-and-socket mechanism developed originally for hip prostheses by Sir John Charnley in the 1960s (Büttner-Janz 2003). The principle underlying this device, substituting the hip articulation with a prosthetic implant, is one of the most successful surgical inventions of recent times - confirmed by its widespread adoption for the treatment of other articulations such as knee and shoulder. These developments were pivotal for bolstering both orthopaedic surgery and the implants industry (Miller 2002). The SB Charité artificial disc was the first hip-like design used in humans.

Hip-like artificial discs feature rigid contact surfaces in the form of a ball-and-socket articulation made of material akin to that used in hip prostheses, i.e. metal or relatively rigid plastic such as UHMWPE (Figure 4.3). Although these discs enhance the mobility of the intervertebral segment, their rigid surfaces prevent effective load absorption (Le Huec *et al.* 2003).

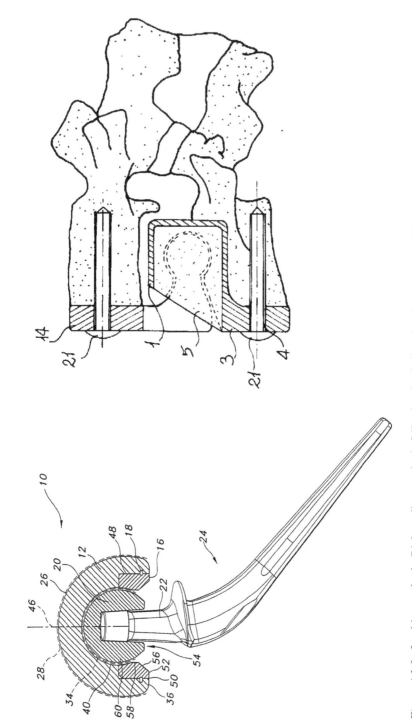

Figure 4.3 Left: a hip prosthesis. Right: a disc prosthesis following the 'ball-and-socket' principle of hip implants.

Source: US6986792 and US5755796.

The alternative to the 'hip-like' operational principle is the 'mimetic', adopted first in Europe in 2007 while still awaiting approval in the US. Although mimetic discs are late entrants in clinical use, since the early 1980s several R&D projects (such as the Acroflex project) have been based on designs following this operational principle, especially in the US (Szpalski *et al.* 2002). Mimetic-type artificial discs are designed to replicate the articulation of the anatomical disc in relation not only to movement but also load absorption (Figure 4.4). However, as already mentioned, since the effective impact of load absorption for both natural and artificial discs has not been established with certainty, opinions concerning these operational principles diverge. Advocates of the hip-like disc maintain that the absorption of load (if it exists) in the anatomical disc is irrelevant and that prosthetic restoration of movement suffices (Mayer 2005). Conversely, advocates of the mimetic disc argue that disregarding load absorption generates systemic biomechanical problems and painful symptoms that often end up requiring further surgery to adjacent levels (Lee and Goel 2004)[6].

In addition to this basic uncertainty about load absorption, there is a fundamental uncertainty about the differential clinical performance of the mimetic and the hip-like discs. To our knowledge, no clinical or laboratory study of any type has been conducted to compare the performance of the two artificial disc principles. This is possibly because specialists are preoccupied by an even bigger task, i.e. proving the efficacy and safety of the whole artificial disc rationale as compared with the dominant technique of spinal fusion. History shows that this task has proved to be difficult and that the randomised studies carried out so far provide no conclusive evidence in either direction.

Thus, there are two fundamental uncertainties related to testing. The first stems from the difficulty involved in measuring load absorption in a disc – whether healthy, diseased or artificial. The second is the lack of tests aimed at measuring the comparative performance of different operational principles of disc

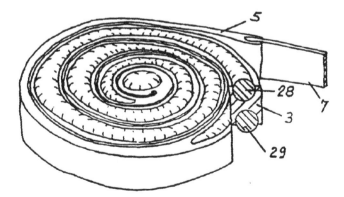

Figure 4.4 A 'mimetic design' based on the reproduction of the viscoelastic properties of the anatomical disc, using materials such as synthetic elastomers.

Source: US6610094.

prosthesis. The contested terrain here is the comparison with spinal fusion that continues to attract resources and attention.

These two fundamental uncertainties triggered different responses concerning the proposed variety of artificial designs. The first response concerned the basic uncertainty over load absorption. This promoted mimetic/hip-like type since uncertainty about the concrete amount of load absorption is behind the decision to incorporate this property into the configuration of the device (mimetic) or not (hip-like design). The second response adds the hybrid design to the mimetic/hip-like repertoire, and is related to testing uncertainty about the clinical differential performance of these two competing principles. Section 3 discusses whether this uncertainty is crucial for the rationale of the hybrid design.

Hybridisation and the artificial disc

This section outlines the basic design principles and the steps that led to the definition of the Bryan artificial disc, the only hybrid device that has been approved for clinical use.[7] We begin by noting that the dualism between the hip-like and mimetic operational principles is a recurrent theme in the specialist medical literature. A survey article by Bono and Garfin (2004) refers to 'articulated non-elastic discs' and 'elastic discs (with load absorption)'. Also Lee and Goel (2004) refer to 'kinematic discs' and 'kinematic and absorption of load discs' while Szpalski *et al.* (2002) discuss 'artifacts destined to restore the kinematic functions' and 'to restore the viscoelastic functions' respectively. Interestingly only the latter study acknowledges that the history of the artificial disc features 'some devices [that] attempt to combine both principles' (Szpalski *et al.* 2002, p. S67). Hints at hybridisation are contained in patent no. US5314477, which refers to the possibility of a 'combination of these two research routes' in the design of the artificial disc, and in patent no. US7563286 whose classification of disc devices includes a hybrid category with different design principles.

To appreciate the hybridisation trajectory we examine the circumstances underpinning development of the Bryan artificial disc. The clinical origins of this venture are discussed in a published interview with Dr Bryan, a spine surgeon (Mutilescu 2002, p. 8):

> In North West there is a very young population which is very active in outdoor activities, whether mountain climbing, topping trees, fishing and so on. Many young people present to us with … early degenerative changes in their spine and get operated with the usual operating procedure. Then these people are coming to us after 5/6 years and in a significant number of them, we find ourselves re-operating… So I have been thinking of this for many years.

To obtain engineering advice, Bryan joined forces with Alex Kuntzle, a mechanical engineer from the metallurgy industry. Following a patent application in 1994 they sought venture capital to fund Spinal Dynamics, a start-up to develop a spinal disc device. The company was subsequently acquired by Medtronic for

$269.5 million at the beginning of the 2000s. The Bryan disc received approval for clinical use in Europe in 2001 and in the US in 2009 (Biondo and Lown 2004; FDA 2009). It has been estimated that, pre-2010, the Bryan disc was used in some 35,000 surgeries. Other artificial disc projects launched during the 1990s followed a similar trajectory: after initial development by a small company with the strong involvement of one or several surgeons, the designs were sold to one of the big five US-based companies that dominate the orthopaedic industry. The two companies that have led the development of the hip-like and mimetic artificial discs (Acromed and Link) were ultimately bought by Johnson & Johnson. This trend is dominant also in the broader spinal sector (Biondo and Lown 2004).

Dr Bryan claimed that the thrust of the hybrid disc venture was to seek 'to change the nature of the joint from an arthrodial joint to restore something similar to diarthrodial joint' (Mutilescu 2002, p. 8).[8] While this statement would seem to refer to the hip-like principle, Dr Bryan claimed that the artificial disc would 'cushion as the normal vertebral disc' (Mutilescu 2002, p. 8), this 'cushioning' being a distinctive property of mimetic design. Patent no. US7025787 (authored by Bryan, Kuntzle *et al.* and owned by Medtronic) claims that the implant 'should also provide elasticity and damping sufficient to absorb shocks and stresses imposed on it in a manner similar to that of the natural disc'.[9] Patent no. 7025787 (2002) describes the salient features of the Bryan artificial hybrid device: one elastomer[10] and two small metal plates that act as the disc-bone interface to ensure stability. These specifications suggest close similarity with the mimetic Acroflex disc (Patent no. US5071437, left side of Figure 4.5).

Early testing of the Acroflex device in both laboratory and human trials highlighted repeated failures of the elastomer - the component expected to provide load absorption and motion - and after several redesigns and test iterations, the project was abandoned (Kurtz 2006). This proved crucial for the conception of the Bryan disc whose layout features an important variation when compared with the basic mimetic configuration: the small plates are allowed to rotate around the elastomer thus creating a ball-and-socket articulation similar to hip-like discs

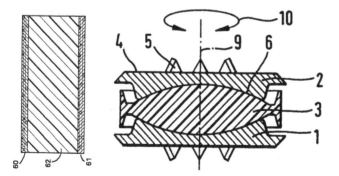

Figure 4.5 Left: a classic elastomer/vertebral plate configuration of a mimetic patent related to the Acroflex project. Right: a typical hip-like configuration.

Source: US5071437.

(right side of Figure 4.5). The mimetic operational principle, in contrast, features small plates joined to the elastomers, in a 'sandwich configuration' that does not allow rotation (left side of Figure 4.5). The existence of elements of both operational principles in the Bryan disc design is also signaled in the citations on the first patent associated with the Bryan disc which include mimetic (as US4911718) and hip-like (DE3023353) patented designs.

It should be noted that the Bryan disc (Figure 4.6) emerged within a context of debate about the performance of the sandwich configuration in the mimetic principle and the positive performance of the new redesigned SB Charité, a hip-like device. The initial bad results of early prototypes of SB Charité in the mid-1980s were counterbalanced in the 1990s by the first good clinical results from a redesigned device (Griffith *et al.* 1994). Almost contemporaneously, the first negative results of experimental trials with the Acroflex mimetic disc were published (Enker *et al.* 1993).

We now provide a functional analysis of the technical aspects involved in the hybrid disc evolution. Figure 4.7 depicts the design trajectories of the hip-like, mimetic and hybrid principles. To highlight their design characteristics we refer to what Ulrich (1995) calls the 'iota level', viz. the individual pieces.

Pioneer prototypes of the artificial disc featured steel (Fenström; Reitz and Joubert) or silicone spheres (Nachemson; Fassio and Ginestie) aimed at improving mobility and load absorption (the silicone spheres) or mobility only (the stainless steel devices). It soon became clear that the spheres were not suited to reproducing the properties of cylindrical discs to support adjacent vertebrae

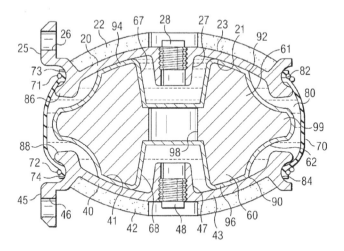

Figure 4.6 Element 99 functions to absorb load, as in the mimetic operational principle patents. However, instead of being joined to the small plates 20 and 40, they move over it, as in hip-like patents. Element 70 is a membrane designed to avoid migration of particles from the movement of the articulation between 99 and 20 and 40.

Source: US7025787.

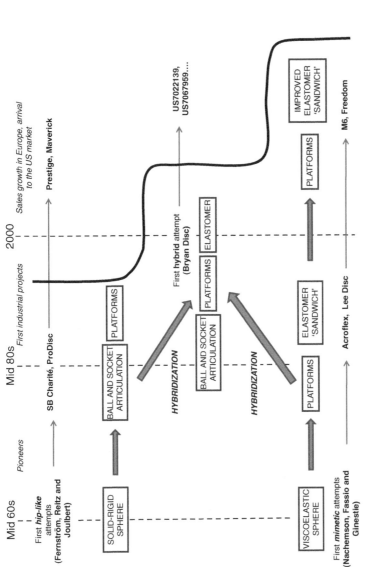

Figure 4.7 The evolution of the artificial disc. The thick black line marks the emergence of each operational principle in clinical use.

Source: own elaboration of Szpalski *et al.* 2002; Lee and Goel 2004; Kurtz 2006; Bono and Garfin 2004; Pimenta *et al.* 2010; Biondo and Lown 2004.

(Figure 4.8). Bono and Garfin (2004, p. 147S) refer to those failures as lessons learned for future developments. One of the efforts in this problem sequence entailed the insertion of two intervertebral plates between the ball and the adjacent vertebrae to improve stability (Bono and Garfin 2004; Lee and Goel 2004). In this second generation of hip-like artificial discs, the ball-and-socket articulation was the component that afforded mobility to the intervertebral space. Subsequent problems guided the incremental transformations to the design to improve the performance of the ball-and-socket articulation (labeled 'Prestige' and 'Maverick' in upper part of Figure 4.7). In the mimetic principle (lower part of Figure 4.7) two additional plates were attached to an intermediate elastomer to create the sandwich configuration that would provide both mobility and load absorption. However, the elastic layer impaired the stability of the implant, demonstrated by repeated and unexpected failures in clinical testing of the Acroflex disc and this provoked general distrust of this configuration.

As mentioned earlier, in the mimetic design the elastomer and the plate are joined in the same mould, a feature that disregards relative rotation. This constraint creates excessive load on the elastomer with the risk of structural failure of the device (Fraser *et al.* 2004). The hybrid model that appeared in the mid-1990s (central part of Figure 4.7) combines the ball-and-socket and elastomer elements in a single device, thus providing mobility and load absorption as in the mimetic principle but in a different way, since the vertebral plates are not joined to the elastomer but articulated in a ball-and-socket joint. Thus the mimetic type failures of the elastomer were overcome by the hybrid Bryan disc design.

Hybridisation of the artificial disc entailed a number of incremental ameliorative efforts. Part 70 in Figure 4.6 indicates a membrane attached to vertebral plates whose function is to prevent the migration of plastic particles from the hip-like articulation. This was important since these particles produced allergic reactions and resulted in the systematic failure of hip implants. The adoption of special plastic support (UHMWPE) addressed this particular problem and consolidated hip substitution into one of the most successful surgical procedures worldwide. However, UHMWPE is a rigid plastic incapable of providing significant load absorption (Le Huec *et al.* 2003). To accommodate this operating principle the Bryan hybrid design needed a more elastic material. For one candidate, polyurethane, there were concerns over the biocompatibility of the particles, which it

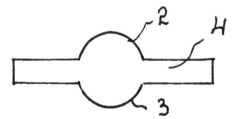

Figure 4.8 The silicon sphere patented by Fassio.

Source: FR2372622.

was expected would produce a negative response when compared with UHMWPE (Naidu 2007). The Bryan design was modified by the inclusion of an isolating membrane that avoided the deleterious effects of using polyurethane (Figure 4.9). Clearly the extra design efforts increased the complexity of the device that, according to other studies of hybridisation (for example, the hybrid car), is inevitable (Dijk and Yarime 2010).

The Bryan disc entered the European market in 2001[11] followed soon after by several hip-like discs. Sales data reported in Biondo and Lown (2004) show that in 2004 the Bryan disc dominated two-thirds of the cervical disc market (2,500 surgeries), with the remaining third shared among the other three hip-like designs (1,300)[12]. It is estimated that in 2013 35,000 surgeries have been performed to implant the Bryan disc, mostly in Europe. Although no production cost data are available it can be expected that the Bryan hybrid device is not any cheaper than other hip-like discs. Indeed, the preceding paragraphs suggest the opposite due to the complexity of the Bryan disc whose design entails more components and assemblies when compared with the 'good old' technologies (Henningfeld 2005, p. 2) based on the articulation concept of hip prosthesis (Miller 2002). This would seem to confirm that the surgeon is the ultimate decision maker and is influenced strongly by product quality. Lieberman (2004, p. 610) points out that in the spinal implants market 'price has become a secondary determinant of demand, only remotely involved because physicians, who function as the principle users, are not the typical final consumer'.

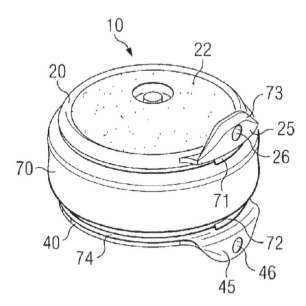

Figure 4.9 The Bryan artificial disc showing the metallic vertebral plates (element 22) and the external membrane (element 70).

Source: US7025787.

We can summarise the technical configurations discussed so far. Hybridisation is a design strategy aimed at embodying all the most useful operational principles, in order to counteract the lack of comparative performance testing, by bundling them together in a single artifact regardless of whether they compete functionally. We discussed how the hybrid Bryan device combines the known positive characteristics of the hip-like disc with the additional potential advantage of coping with load absorption. Table 4.1 compares the advantages and disadvantages of the three principles discussed so far.

Conclusions

The analysis presented here illustrates the transition in inventive activity that led to the definition of the hybrid Bryan disc in the mid-1990s. We analysed the 1973–2004 period, which was an intense period of innovation in the ongoing period of development. The technological trajectory analysed is still unfolding and many other potential developments are possible.

The history of the artificial disc shows a series of shifts in the more recent years, related to the dominant anatomic approaches. The original designs for lumbar prostheses were conceived in the 1980s. In the 1990s the first cervical artificial discs were developed. More recently, the cervical approach has achieved greater market penetration than the lumbar solutions where doubts still exist about the advantages of lumbar arthroplasty over lumbar fusion (Fekete and Porchet 2010; see also note 7). The most recent results of cervical arthroplasty are more conclusive (Blumenthal *et al.* 2013). These anatomic differences would seem to be caused by the more demanding biomechanical requirements of the lumbar spine (Kurtz 2006). Regarding the mimetic/hip-like design dichotomy, since 2010 there has been an increase in the adoption of mimetic designs in Europe and they are under pre-approval trials in the US (lower-right side of Figure 4.7), for both cervical and lumbar designs. If the resurgence of the mimetic design finally confirms the clinical superiority of this principle, it would be reasonable to suppose that the hybrid disc should be competitive since its configuration should also allow load absorption.

The case study of the artificial disc illustrates the extent to which design heuristics contribute to devising a remedy for a practical problem, back pain, in the absence of definitive knowledge on the nature of the cause (Degenerative Disc Disease) and the comparative performance of the devices designed to alleviate its painful effects. More specifically, we showed that hybridisation - the embodiment of multiple competing operational principles within a single medical device - is a strategy for dealing with the practical shortcomings stemming from uncertainty about the comparative clinical performance of available therapeutic solutions. Looking at this process from a historical perspective reveals that the long-term trajectory of medical innovation does not resemble a punctuated succession of alternative paradigms but rather incremental inter-penetration of different problem-solving styles in which emergent courses co-exist with, rather than replace, existing ones (Pickstone 1993).

Table 4.1 Summary of artificial disc designs.

	Design elements	Services	Comparative advantages	Comparative disadvantages
Mimetic	Sandwich configuration, with a constrained elastomer.	Motion and a presumed amount of load absorption.	A presumed amount of load absorption.	(Early 1990s.) Negative experimental clinical results of the Acroflex disc, with systematic failure of the elastomer in the sandwich configuration.
Hip-like	Ball-and-socket articulation between rigid materials.	Motion.	(Early 1990s.) Positive clinical results of the SB Charité and Frenchay Hospital discs. The ball-and-socket configuration avoids the constraints of the elastomer typical of mimetic designs.	No load absorption.
Hybrid	Ball-and-socket articulation between elastomer and two metallic plates.	Motion and a presumed amount of load absorption.	(Invented in the mid-1990s.) A presumed amount of load absorption. The ball-and-socket configuration avoids the constraints of the elastomer typical of the sandwich configuration in mimetic designs.	

Acknowledgement

This chapter contains material reprinted from Barberá-Tomás, D. and Consoli, D. (2012) 'Whatever works: uncertainty and technological hybrids in medical innovation'. *Technological Forecasting and Social Change*, 79(5). Copyright 2012, with permission from Elsevier.

Notes

1 According to Murmann and Frenken (2006, p. 939), the concept of operational principles facilitates the systematisation of a set of artifacts into general product classes on the basis of 'the kind of knowledge a human designer must have in order to build a technological device that works on the physical world in a desired way'.

2 The physiology of the DDD in the lumbar and the cervical areas is very similar, and so are the prostheses.

3 In addition to the biomechanical explanation there is a purely chemical explanation that refers to the pain-provoking chemical changes of the discs during the degeneration process (Bono and Garfin 2004).

4 An extreme version of this was proposed by Mulholland (2008) who suggested that loads are the only biomechanical cause of pain in the disc, and that movement-related causes are 'a myth' that has dominated spinal biomechanics for the last 30 years.

5 In the words of an artificial disc designer: 'You cannot put a sensor in an *in-vivo* environment to measure load absorption. And that means that, at best, you can only have *in-vitro* data for this property in the healthy, degenerated and artificial disc. And you cannot trust only *in-vitro* data since these are highly dependent on the specific design of the laboratory experiment'. Furthermore, laboratory experiments simply cannot reproduce some conditions of the *in-vivo* environment: 'The dynamic load-response behavior of the anatomic disc could depend even of the hour of the day, and you cannot reproduce that with cadaveric specimens'.

6 Note that the argument related to the systemic problems generated in adjacent levels has also been presented as the most important advantages of the artificial disc over osseous fusion or arthrodesis. In this context, it is the abnormal *motion* caused by osseous fusion that is claimed to cause the biomechanical problems in adjacent levels. What the discussion on mimetic and hip-like discs adds to this debate is that also natural *load absorption* has to be restored to maintain the biomechanical equilibrium of adjacent levels.

7 The technical specifications presented were validated through interviews with four experts, one of whom was involved in the development of the Bryan disc in the late 1990s, another was a design engineer on one of the most important mimetic projects in the late 1980s - the Acroflex artificial disc. The other two experts are R&D engineers currently working in new artificial disc developments, one hip-like and the other mimetic.

8 Diarthrodial joints (or synovial joints), such as hip or knee, joints are freely moveable; arthrodial joints (or cartilaginous joint), as the spinal disc, allow only limited movement.

9 The patent also states that the objective of the invention is to transform a natural arthrodial joint such as the spinal disc into an artificial diarthrodial joint.

10 Elastomer refers to materials with mechanical properties (for example, hysteresis) similar to rubber.

11 Artificial discs are used in lumbar and cervical surgery. Although the sizes of the artifacts depend on the spinal area, patented designs usually refer to the spinal zone, with no distinction between cervical and lumbar prostheses. In general, development of disc prostheses depends strongly on the capabilities of the surgeons involved.

Bryan, for example, was mainly concerned with the cervical region. At the same time the technological principles of the Bryan disc are applicable to both the cervical and lumbar regions: in patents the artifact is referred to as a 'spinal disc endoprosthesis', without specifying the anatomical region involved.

12 The Bryan disc was ranked second behind the lumbar SB Charité, in a list of 9 lumbar/cervical artificial discs sold worldwide.

References

Biondo, D. and Lown, D. (2004) *Beyond Total Disc. The Future of Spine Surgery*, Spine Industry Analysis Series. New York: Viscogliosi Bros LLC

Blumenthal, S. L., Ohnmeiss, D. D., Guyer, R. D. and Zigler, J. E. (2013) 'Reoperations in cervical total disc replacement compared with anterior cervical fusion: results compiled from multiple prospective Food and Drug Administration investigational device exemption trials conducted at a single site'. *Spine*, 15; 38(14), pp. 1177–82

Bono, C. M. and Garfin, S. R. (2004) 'History and evolution of disc replacement'. *The Spine Journal*, 4, pp. 145–150

Boos, N. and Webb, J. (1997) 'Pedicle screw fixation in spinal disorders: a European view'. *European Spine Journal*, 6, pp. 2–18

Büttner-Janz, K. (2003) 'History' in K. Büttner-Janz, S. Hochschuler and P. McAfee (Eds) *The Artificial Disc*. Berlin: Springer, pp. 1–10

Cummins, B., Robertson, J. and Gill, S. (1998) 'Surgical experience with an implanted artificial cervical joint'. *Journal of Neurosurgery*, 88, pp. 943–948

Dahl, M., Rouleau, J., Papadopoulos, S., Nuckley, D. and Ching, R. (2006) 'Dynamic characteristics of the intact, fused, and prosthetic-replaced cervical disk'. *Journal of Biomechanical Engineering*, 128(6), pp. 809–14

Dijk, M. and Yarime, M. (2010) 'The emergence of hybrid-electric cars: innovation path creation through co-evolution of supply and demand'. *Technol. Forecasting and Social Change* 77 (8), pp. 1371–1390

Duke University Medical Center (2004) 'Economic impact of back pain substantial'. *ScienceDaily*

Enker, P., Steffee, A., McMillin, C., Keppler, L., Biscup, R. and Miller, S. (1993) 'Artificial disc replacement. Preliminary report with a 3-year minimum follow-up'. *Spine*, 18(8), pp. 1061–1070

Errico, T. J. (2005) 'Lumbar disc arthroplasty'. *Clinical Orthopaedics & Related Research* 435, pp. 106–117

Food and Drug Administration (2009) 'Bryan disc approval letter'. Available online at: www.accessdata.fda.gov/cdrh_docs/pdf6/P060023a.pdf (retrieved 21 April 2010)

Fraser, R. D., Ross, E. R., Lowery, G. L. and Freeman, B. J. (2004) 'Lumbar disc replacement. AcroFlex design and results'. *The Spine Journal*, 4, pp. 245S-251S

Freeman, B. and Davenport, J. (2006) 'Total disc replacement in the lumbar spine: a systematic review of the literature'. *Euro Spine Journal*, 15 (3), pp. S439–S447

Fekete, T. F. and Porchet, F. (2010) 'Overview of disc arthroplasty - past, present and future'. *Acta Neurochir*, 152, p. 393

Geisler, F. (2006) 'The Charite artificial disc: design history, FDA IDE study results and surgical technique'. *Clinical Neurosurgery*, 53, pp. 223–228

Griffith, S. L., Shelokov, A. P., Shelokov, A. P., Buttner-Janz, K., LeMaire, J. P. and Zeegers, W. S. (1994) 'A multicenter retrospective study of the clinical results of the LINK SB Charite intervertebral prosthesis: the initial European experience'. *Spine* 19, pp. 1842–1849

Henningfeld, C. (2005) 'Executive interview'. *Orthoknow*, July, pp. 1–3

Kurtz, S. M. (2006) 'Total disc arthroplasty' in S. M. Kurtz and A. A. Edidin (Eds) *Spine Technology Handbook*. New York: Academic Press

Le Huec, J. C., Kiaer, T. and Friesem, T. (2003) 'Shock absorption in lumbar disc prosthesis: a preliminary mechanical study'. *Journal of Spinal Disorders & Techniques*, 28, pp. 346–51

Lee, C. and Goel, V. (2004) 'Artificial disc prosthesis: design concepts and criteria'. *The Spine Journal* 4(6), pp. S209–18

Lieberman, I. (2004) 'Disc bulge bubble: spine economics 101'. *The Spine Journal* 4(6), pp. 609–613

Maniadakis, A. and Gray, A. (2000) 'The economic burden of back pain in the UK'. *Pain*, 84, pp. 95–103

Mayer, H. M. (2005) 'Total lumbar disc replacement'. *J Bone Joint Surg BR*, 87, pp. 1029–37

McAfee, P. (2003) 'An explanation of early, suboptimal results from Charité Hospital' in K. Büttner-Janz, S. Hochschuler and P. McAfee (Eds) *The Artificial Disc*. Berlin: Springer, pp. 1–10

Medical Device and Diagnostic Industry News (MDDI) (2008) 'Spine sector shows continued growth'. Available online at: www.mddionline.com/article/spine-sector-shows-continued-growth (retrieved 11 December 2012)

Metcalfe, J. S., Mina, A. and James, A. (2005) 'Emergent innovation systems and the development of the intraocular lens'. *Research Policy*, 34, pp. 1283–1304

Miller, D. (2002) 'Orthopaedic product technology during the second part of the twentieth century' in L. Klenerman (Ed.) *The Evolution of Orthopaedic Surgery*. London: Royal Society of Medicine Press

Mina, A., Ramlogan, R., Tampubolon, G. and Metcalfe, J. S. (2007) 'Mapping evolutionary trajectories: applications to the growth and transformation of medical knowledge'. *Research Policy* 36(5), pp. 789–806

Morlacchi, P. and Nelson, R. R. (2011) 'How medical practice evolves: learning to treat failing hearts with an implantable device. *Research Policy*, 40(4), pp. 511–525

Mulholland, R. C. (2008) 'The myth of lumbar instability: the importance of abnormal loading as a cause of low back pain'. *Euro Spine Journal*, 17, pp. 619–625

Murmann, J. P. and Frenken, K. (2006) 'Toward a systematic framework for research on dominant designs, technological innovations, and industrial change'. *Research Policy*, 35(7), pp. 925–52

Mutilescu, A. (2002) 'Interview with Dr Vincent Bryan'. *Argos Spine News*, 6(Oct.), pp. 7–11

Naidu, S. (2007) Comments in the *Summary Minutes of the Orthopaedic And Rehabilitation Devices Panel Meeting*. July 17, Washington DC

Pickstone, J. V. (1993) 'Ways of knowing: towards a historical sociology of science, technology and medicine'. *The British Journal for the History of Science*, 26, pp. 433–458

Pimenta, L., Springmuller, R., Lee, C. K., Oliveira, L., Roth, S. E. and Ogilvie, W. F. (2010) 'Clinical performance of an elastomeric lumbar disc replacement: minimum 12 months follow-up'. *SAS Journal*, 4 pp. 16–25

Singh, K., Vaccaro, A. and Albert, T. (2004) 'Assessing the potential impact of total disc arthroplasty on surgeon practice patterns in North America'. *The Spine Journal*, 4(6), pp. S195-S201

Szpalski, M., Gunzburg, R. and Mayer, M. (2002) 'Spine arthroplasty: a historical review'. *Euro Spine Journal*, Suppl 2, pp. S65–84

Ulrich, K. (1995) 'The role of product architecture in the manufacturing firm'. *Research Policy*, 24, pp. 419–441

Urbaniak, J. R., Bright, D. S. and Hopkins, J. E. (1973) 'Replacement of intervertebral discs in chimpanzees by silicondacron implants: a preliminary report'. *Journal of Biomedical Material Research* 7, pp. 165–86

Vincenti, W. (1990) *What Engineers Know and How They Know It: Analytical Studies from Aeronautical History*. Baltimore: Johns Hopkins University Press

5 Technological accretion in diagnostics

HPV testing and cytology in cervical cancer screening

Stuart Hogarth, Michael Hopkins and Daniele Rotolo

Introduction

This chapter follows the emergence of molecular HPV testing technologies and their application to cervical cancer screening in the USA. When HPV testing was first commercialised in the late 1980s, screening for cervical cancer had been a routine part of preventive healthcare for at least two decades, with testing conducted by cervical cytologists using the Pap smear test, a technology first developed in the 1910s. Many now predict that molecular HPV tests will eventually replace cervical cytology, bringing fundamental changes to the clinical infrastructure of screening in the process. However, at present these technologies not only co-exist but augment each other. This process of technological accretion has involved not only an accommodation with molecular technologies, but also the widespread adoption of novel cytology technologies: liquid-based cytology (LBC) and automated slide readers. This chapter explores the institutional factors that have shaped this contingent outcome. We begin by setting out the clinical context of cervical cancer screening.

Cervical carcinoma is the fourth most common form of cancer in women, and is the cause of 7.5 per cent of cancer deaths in women worldwide (IARC 2012). These global statistics belie a grossly unequal disease burden: cervical cancer is now predominantly a disease of low- and middle-income countries because, since the 1960s, both incidence and mortality rates have dropped dramatically in many developed countries. This is generally ascribed in large part to the introduction of screening programmes. Cervical cancer screening (CCS) has become ubiquitous in the developed world with many countries running national programmes to ensure that women have the opportunity for regular screening. Statistics from the USA show a greater than 50 per cent decline over 30 years in both incidence and mortality, which are widely attributed to cervical cancer screening, although the disease still kills around 2.38 in 100,000 women in the USA (ACOG 2012). The decline in incidence and mortality is due primarily to that fact that screening identifies pre-invasive lesions that can be treated well before the possible onset of cancer (Saslow *et al.* 2012).

Until recently cytology-based screening has relied on a technology developed in the first half of the twentieth century: the Pap smear. The traditional Pap smear involves scraping cells from the cervix and smearing them in a thin layer on a glass slide. The cells are then stained and examined under a microscope by a cytologist to check for abnormalities. In the 1990s the traditional Pap began to be replaced by liquid-based cytology – a technique that involves placing the cells into preserving fluid and then filtering them to remove impurities prior to examination by microscope. However, even before LBC became routine, a more radical alternative technology was in development. In 1983 scientists provided strong evidence for an association between cervical cancer and human papilloma virus (HPV) when they cloned two carcinogenic HPV types (HPV 16 and 18) and, soon thereafter, companies began to develop HPV tests for use in cervical cancer screening.

Method

Developing a robust history of HPV testing in the USA required a mixed method historical process study that triangulates several data sources (Van de Ven 2007). First we undertook exploratory searches of trade and scientific literature and diagnostics industry news websites to reveal key investors and organisations in the HPV field. Bibliographic searches aided detailed mapping of these organisations, their activities and links. We collected publication data by querying the Thomson-Reuter's Web of Science (WoS) database using an *ad hoc* search string to identify those records related to HPV diagnostics.[1] This returned a sample of 1,560 scientific articles published up to 2011. Figure 5.1 depicts the number of articles related HPV diagnostics over time. The most active organisations in

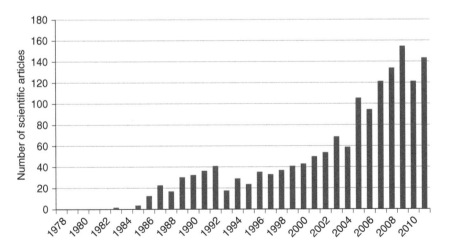

Figure 5.1 Published scientific articles related to HPV diagnostics (up to 2011).

Source: authors' elaboration on the basis of ISI Web of Science data.

Table 5.1 Top-5 profit/non-profit organisations in terms of number of published scientific articles

Non-profit organisation	Number of publications	Profit organisation	Number of publications
1) National Cancer Institute (NCI)	65	1) Digene (Qiagen since 2006)	35
2) Johns Hopkins University	51	2) DDL Lab	9
3) University Vrije Amsterdam	49	3) Roche	9
4) University of Turku	32	4) Abbot	5
5) McGill University	30	5) Cetus	5

Source: authors' elaboration on the basis of ISI Web of Science data.

terms of published scientific articles are reported in Table 5.1. We distinguished these organisations as non-profit and profit.[2]

Data on patents related to HPV testing were also collected.[3] Using these data, interviewees were selected based on their involvement with developments in HPV diagnostics in the USA and EU as evidenced by authorship of major papers, patents and guidelines. Industry executives, clinical scientists, laboratory directors and physicians were interviewed, including supporters of technological options that compete with Digene's kits. A pilot study in 2008 involved 12 interviews with figures from the EU and USA; in 2013–14 this was supplemented by a further 26 interviews in the USA with industry executives, clinician-researchers, regulatory officials and government scientists. Semi-structured interviews of 40–150 minutes were recorded and fully transcribed. Interviews have well-documented limitations as sources for recent histories (Hughes 1997). In particular, social scientists need to be reflexive about interviewees' partisan nature and should triangulate data from different sources (Van de Ven 2007). We therefore used interviews selectively, primarily to identify key themes and events, and we also used multiple industry interviewees to triangulate view points within and between firms with a major role in the history.

The development of cervical cancer screening in the USA

To put the adoption of HPV testing in context, it is important to understand the origin of CCS, the Pap test and the potential niches that HPV testing might occupy.

The emergence of early detection as a public health goal

The annual Pap smear test has made cervical cancer screening a cornerstone of preventive health practices in the USA. However, women were being encouraged to have regular pelvic examinations decades before the Pap entered clinical practice. At the beginning of the twentieth century senior cancer surgeons became convinced that the earlier cancer could be treated, then the greater the likelihood that treatment might be successful (Lowy 2011). In 1913 doctors founded the American Society for the Control of Cancer (ASCC), with the primary goal to promote the benefits of early detection to physicians and the public (Gardner 2006, pp. 20–25). Although campaigners were promoting a generalised message of cancer control, the examples used were generally female cancers – breast or uterus. By the 1930s the importance of annual vaginal examinations was being promoted by women's organisations such as the YWCA, who equipped a number of health centres with facilities to conduct examinations. During the 1930s and '40s government agencies such as the National Cancer Institute and the US Public Health Service took up the cause of early intervention. It was in this context that new screening tools came to be of interest (Casper and Clarke 1998; Lowy 2011).

The development of the Pap smear

George Papanicolaou developed the Pap smear in the Department of Anatomy at Cornell University, initially through research on the menstrual cycle of guinea pigs. In 1917 he published a paper demonstrating that the microscopic examination of cervical cells collected by his vaginal smear method could be used to identify stages of the oestrus cycle. Papanicolaou then turned his attention to humans. Having demonstrated the applicability of the technique as a metric of the oestrus cycle in women, he began collecting specimens from women suffering various gynaecological conditions. Papanicolaou discovered that nearly all the specimens from cervical cancer cases contained cells that could be identified as abnormal. He published his findings in 1928, but was so discouraged by the negative response of the pathology community that a decade passed before Papanicolaou resumed work applying his smear technique to cancer detection, this time in collaboration with Herbert Traut, a pathologist at Cornell's gynaecology department. They published their first paper in 1941 and, by 1948, the technique was gaining institutional support: the National Institutes of Health began funding large-scale studies on cervical cancer screening using the Pap smear, and the American Cancer Society and the US Public Health Service financed training courses for pathologists. The process of pathology professionalisation was given institutional structure in 1951 when the Inter-Society Cytology Council (ISCC) was established. By the 1960s the Pap smear was widely available to women in the USA.

The Pap smear exemplifies a technology both deeply entrenched in medical practice and highly unstable; from its earliest clinical adoption in the 1940s, the Pap was subject to repeated modification, including changes to the labour process, the technology, the classification systems and the governance regimes that ensure its safety and effectiveness. Casper and Clarke's (1998) seminal account of Pap testing in the USA described how it has had to be 'massaged and manipulated' to transform it into a reasonably 'right tool' for cervical cancer screening. Their paper was an explanation of what we might term the Pap paradox: the test is widely credited with lowering cervical cancer mortality internationally, and has been

Table 5.2 Three possible roles for HPV testing in cervical cancer screening

Testing protocol	Description
ASC-US triage (reflex)	Cytology remains screening test, HPV used only for follow-up in case of ambiguous smear test (ASC-US), reducing need for colposcopy.
Co-testing	Cytology and HPV used as joint screening tests, allowing less frequent screening for women who test negative for both tests.
HPV screening	HPV used as screening test, cytology used as a follow-up for HPV-positive women.

Source: developed by authors from primary material.

described as 'the most effective screening test for cancer that has ever been devised' (Dehn *et al.* 2007). But, with 15–50 per cent false-negative rates (i.e. failure to identify cervical cancer when it is present), the Pap has long been problematised as expensive, subjective and error-prone (Cox and Cuzick 2006).

The visual acumen of cervical cytologists remains central to cervical cancer screening, despite protracted and expensive attempts to replace or automate the process (Casper and Clark 1998; Keating and Cambrosio 2003). Alternative methods include screening colposcopy, visual inspection with acetic acid (VIA) or with Lugol's iodine (VILI), real-time imaging and tumour markers - yet it is cytology that is synonymous with cervical cancer screening. However, this chapter focuses on the mounting pressure on cytology from a new molecular diagnostic technology: DNA tests for Human Papillomavirus (HPV) infections associated with the onset of cervical cancer.

HPV testing has already moved from its initial niche as a triage test for ambiguous cytology results, to a more central role in co-testing (i.e. used together with cytology), but in the future it could supplant the established technology (see Table 5.2). To understand the technological trajectory of the HPV test, we also chart how technological innovation *within* cervical cytology has maintained the dominance of the entrenched socio-technical regime.

The career of the HPV test

The dynamics of technological innovation can be followed longitudinally using Blume's concept of the *career*, a sequence of milestones and phases that are (i) exploration, (ii) development, (iii) adoption, and (iv) growth (Blume 1992). Building on this approach we identified the key events that shaped the emergence of HPV diagnostics and clustered them according to Blume's four phases (see Table 5.3).

Exploration (1983–1990)

In the 1980s Professor Harald zur Hausen's team at the German Cancer Research Centre discovered an association between HPV type 16 and cervical cancer

Table 5.3 Key events in the emergence of the HPV testing in the USA

Phase	Period	Event
Exploration	1972-1983	zur Hausen and Meisels separately hypothesise association between the HPV infections and cervical cancers. zur Hausen and colleagues clone HPV 16 in 1983.
	mid-1980s	BRL-Life Technologies (BRL-LT), in collaboration with Georgetown University, begin development of a commercial HPV test. Attila Lorincz (BRL-LT) and George Roth (Institut Pasteur) discover novel oncogenic HPV types which they then patent.
	1988	BRL-LT becomes the first company to gain the FDA approval for an HPV test: Virapap.
	1989	Cetus patent HPV PCR primer set.
	1990	Lack of clinical uptake leads BRL-LT to sell its diagnostic division to Digene.
	1991	Roche acquire Cetus PCR technology.
	1992	Digene develops and patents a new test called Hybrid Capture (HC). Digene begin to collaborate on a series of clinical studies in collaboration with charities, government departments, universities, and research institutes across the world.
Development	1996	Oncor gains conditional FDA approval for HPV test. FDA approve first Liquid-Based Cytology test (owned by Cytyc). Cytyc and Digene form alliance for development and marketing.
	1998	Ventana acquire Oncor HPV test (and related technologies).
	1999	Digene gains FDA approval for the adoption of its second-generation test (Hybrid Capture II) in ASC-US triage testing protocol.
Adoption	2001	Data from NCI-funded ALTS trials (for which Digene provided supplies free of charge) generates support for view that HPV testing is best option for AS-CUS triage, supported by the American Society for Colposcopy and Cervical Pathology (ASCCP) in new clinical guidelines. Digene is involved in a series of patent litigations against various rivals (Gen-Probe, Roche, Beckman Coulter, and Third Wave). The litigation ends in 2009.
	2002	Cytyc announce intention to acquire Digene but is blocked by Federal Trade Commission. HPV testing is included in new guidelines issued by the American Cancer Society (ACS) as an adjunctive screen in women over 30 (also known as co-testing).
	2003	FDA approve Hybrid Capture II for co-testing.
	2004	Digene rivals receive FDA warning letters for marketing tests without approval.
Growth	2007	Qiagen acquire Digene. Hologic acquire Cytyc.
	2008	Hologic acquire Third Wave and Roche acquire Ventana.
	2009	Hologic become second company to gain FDA approval for HPV test. FDA issue draft guidance on HPV test approval.
	2011	Roche and Gen-Probe gain FDA approval for their HPV tests.
	2012	New guidelines from ACS, ASCCP, ACOG and USPSTF endorse HPV testing.
	2014	Roche gain approval for use of COBAS test as primary screening tool.

Source: authors' elaboration.

(zur Hausen 1987). Their viral model of cervical cancer aetiology would in time be rewarded with a Nobel Prize, but was initially highly controversial.

The origins of this discovery begin in the late nineteenth century, when a correlation was drawn between cervical cancer, marriage and childbirth, but at this time physicians were still working with an aetiological model of cancer as the product of chronic irritation (Lowy 2011, p. 131). After the Second World War this theory was gradually replaced by an aetiology that focused on cell mutation. This new theory supported an alternative explanation of the genesis of cervical cancer: a sexually transmitted infection (STI). Despite some promising research in the early twentieth century the viral model of cancer aetiology was largely neglected until the discovery of links between viruses and some of the rarer cancers in the 1960s (Lowy 2010, pp. 137–8). The search for a viral cause of cervical cancer initially focused on herpes simplex virus and cytomegalovirus but, in 1976, papers by zur Hausen and by Meisels and Fortin identified a link between the disease and papillomavirus. The preceding year the first international papillomavirus meeting was convened by researchers from the Institut Pasteur signalling the beginnings of a new research domain. However, the significant growth of this field awaited confirmation of the viral theory which, in turn, required the development of new molecular technologies for cloning HPV types. HPV research thus became intertwined with the emergence of new technologies based on recombinant DNA and the development of the molecular diagnostics sector.

The first wave of HPV tests

Within 18 months of zur Hausen's discovery, researchers at laboratory-supplier Bethesda Research Laboratories-Life Technologies (BRL-LT) in Maryland began work on a commercial HPV test, the firm's first foray into clinical diagnostics. The BRL-LT team, led by Attila Lorincz, were fortunate to be located close to one of the pioneering HPV research groups based at Georgetown University: Robert Kuman, a gynaecologic oncologist and pathologist; Wayne Lancaster, a molecular biologist and Ben Jensen, a pathologist with training in virology. Initially BRL-LT's goal was to understand the epidemiology of HPV types 16 and 18 in patients, essential data for clinical validation of an HPV test. However, they discovered that many of the cervical cancer samples from Georgetown were not infected with these known high-risk HPV types. Thus Lorincz's team and their collaborators at Georgetown turned to identifying novel HPV types, discovering and cloning a number of high-risk strains. This work established Lorincz as a leading figure in the nascent field of HPV research.

From the outset the commercial development of the HPV test required the creation of inter-organisational links with academics and clinicians enmeshed in the established regime of cytology-based cervical cancer screening. BRL-LT's research identifying new HPV types illustrates how the corporatisation of biomedical research undermines any simple model of basic research as an academic function, and the patenting of HPV strains by both BRL-LT and their academic counterparts illustrates how the commercialisation of biomedical research has become entrenched in the practices and values of public institutions.

BRL-LT were not alone in exploring the commercial potential of HPV testing in the 1980s. Their US rivals included two first-generation biotechnology firms established in the 1970s: Cetus Corporation and Enzo Biochem (the latter formed in 1976, the same year that Bethesda Research Laboratories was established) and a wave of newer companies, including Digene, Oncor and Vysis. At the time, the fledgling molecular diagnostics sector was expanding rapidly and infectious disease tests – viral and bacterial – were the main driver of that growth.

During this period the most significant technological breakthrough in the emergent field of DNA-based diagnostics was the invention of Polymerase Chain Reaction (PCR) in 1983 by Kary Mullis, a scientist based at Cetus. PCR would become a key platform technology for the molecular diagnostics sector, broadly licensed to multiple firms and used both as a research tool and in clinical diagnostics. Within Cetus, PCR was being applied to HPV detection by a researcher named Michelle Manos and in 1989 this technology was patented as the MY09-MY11 primer set (Manos *et al.* 1989). Cetus facilitated adoption of the new technology by the HPV research community, offering training courses and providing free reagents, and PCR was rapidly adopted – the first publication using the technique came out in 1988 (Shibata *et al.* 1988) and in the same year a number of other scientists were presenting research findings based on their use of PCR techniques at the seventh International Papillomavirus conference.

1988 also marked a major milestone for BRL-LT, as they became the first company to gain FDA approval for an HPV test: the 'ViraPap' kit. The late 1980s might have appeared to have been a propitious time to launch an alternative to the Pap smear. At the time, the Pap smear test was under critical public scrutiny following media revelations that some of the larger laboratories were offering cut-price tests by forcing their staff to read 200 slides per day, more than double the work-rate recommended as safe by the American Society of Cytotechnology (Casper and Clarke 1998). Congress responded by developing new legislation to enforce more rigorous quality assurance mechanisms. However, despite some clinical uptake of ViraPap, regulatory approval proved to be no guarantee of commercial success. This was in part due to the kit's technical limitations - it was radioactive, so had a short shelf life, and was potentially hazardous to lab staff. Furthermore, it was not able to detect a sufficient range of HPV types (of which many were still being discovered). Perhaps more significant than these technical limitations was the lack of clinical data demonstrating that the test benefited patients. Some clinical experts anticipated that adoption was likely to follow once more data was available, but others expressed profound scepticism about the utility of HPV testing (Corliss 1990). The main technological change occurring in cervical cancer screening at this time was not adoption of molecular technology but a more low-tech switch in specimen collection device, from the traditional swabs and spatulas to new endocervical brooms and brush/spatula combinations (Titus 2006). Frustrated at the commercial failure of their foray into cancer screening, BRL-LT sold their molecular diagnostics division for $3.6m to

Digene, a small local rival that had recently been acquired by two entrepreneurs: Charles Fleischman and Evan Jones.

Development (1990–1999)

In a recent historical overview of the HPV research enterprise, Harald zur Hausen summarised the state of scientific knowledge in 1990:

> experimental data indicated already at that time that viral gene expression of latently infected cervical cancer cells is a necessary precondition for growth properties and malignant phenotype of these cells.
>
> (zur Hausen 2006, p.iii.)

However, according to the British scientist Julian Peto, at this stage the views of the HPV research community were not widely shared:

> It was clear to everybody who was interested in HPV by about 1985 or 1986, that this was the cause of cervical cancer. It took us ten years to persuade the rest of the world, and a lot of clinicians were sceptical until quite recently.
>
> (Quoted in Reynolds and Tansey 2009, p.49.)

The doubts raised about the utility of BRL-LT's ViraPap test were thus symptomatic of a broader scepticism at the beginning of the 1990s. As a clinical intervention and as a scientific theory HPV's role in cervical cancer had yet to gain broad acceptance in the USA. However, much would change in this decade. The second phase in the technological trajectory of HPV testing was marked by two key developments: the emergence of an international consensus that high-risk HPV infections, particularly HPV 16 and 18, were a necessary cause of cervical cancer; and growing support for the view that HPV testing could play some role in cervical cancer screening.

Underpinning these developments were two increasingly divergent strands of research: large-scale epidemiologic studies exploring the prevalence of different HPV types and their association with cervical cancer; and clinical studies examining the utility of HPV testing in cervical cancer screening. The clinical research activity was primarily conducted using a new proprietary HPV technology developed by Digene, who emerged in this period as the dominant commercial actor in the embryonic HPV testing market and the only company whose primary focus was HPV testing, while the epidemiological studies generally used PCR. Some of this work was done using the PCR primers developed by Cetus, but the major global research effort led by the International Agency for Research on Cancer chose an alternative set of primers developed in the Netherlands.

This decade saw some of Digene's rivals from the 1980s, such as Oncor and Cetus, fall by the wayside, their assets acquired by other firms. This process of industry churn led to the emergence of new player: Roche Molecular, formed in 1991 as a subdivision of the large Swiss pharmaceutical and diagnostic firm,

Hoffman-la-Roche, when the firm acquired the IP rights to PCR from Cetus (as well as many of the staff from the Cetus PCR group). Michelle Manos, who had led the HPV PCR development at Cetus, did not join Roche Molecular. One reason may have been that the Cetus management who joined Roche did not think HPV was a priority. According to one former Cetus employee, even before the Roche acquisition, Cetus management had taken the view that HPV testing was 'like a fine wine' (USA 25), something that needed time to mature:

> The general consensus was ... we'll let Hybrid Capture spend the money to develop the market and then if that comes out to be something important, we could develop the test quickly and go and take over ....
>
> (USA 25.)

Instead, Roche devoted much of its resources to HIV diagnostics, where there was a rapid growth in viral load testing for prognosis and monitoring of patients. However, the firm retained a presence in HPV research during the 1990s by continuing the Cetus practice of giving scientists free PCR reagents.

Oncor was another potential rival and, unlike Roche, Oncor were committed to HPV testing in the early 1990s, gaining conditional FDA approval for their test in 1996. However, in 1996 Oncor decided that it needed to reconfigure the test 'to facilitate automation and integration of the test with automated Pap Smear testing' (Yuxiang (for Oncor) 1996). Oncor stated that this would entail both a new approval application to the FDA and developing a cooperative agreement with one of the firms manufacturing automated cytology. However, before the firm was able to do this, it ran into multiple difficulties: patent litigation with a rival company forced it to withdraw many of its products from the market and, by October 1998, its financial problems were so severe that trading on its stock was suspended. In an effort to generate cash, it sold its HPV test and related products to Ventana in 1998. Oncor's failure arose in part because of the breadth of its pipeline - much of its R&D activity in the mid-1990s was focused on gaining FDA approval for its Her-2/neu test. According to one former employee the combination of the cost of this process and the loss of revenues arising from the patent litigation meant that Oncor 'didn't really have the wherewithal to do the kind of stuff that ultimately was required to get HPV commercialised' (USA 21). According to a report in the *Washington Post*, Oncor's HPV failure was symptomatic of a broader problem within the company:

> A widely held view among present and former Oncor employees is that the company tried to do too much, too fast, spending its energies and precious capital on a slew of endeavors without focusing on any one of them long enough to make a business out of it.
>
> (Gillis 1998.)

Even at Digene the future of HPV testing (and thus its value as a focal point for the firm's R&D efforts) was the subject of internal debate within the company, a debate that would only be fully resolved in the mid-1990s as the company prepared for its public flotation. The initial focus of the firm was to develop a new platform technology, which might support a range of tests, as Cetus had done with PCR.

Developing the HPV testing technology

The new platform, called Hybrid Capture (HC), was already in development at BRL-LT in the late 1980s, and was patented by Digene in 1992. It was a non-radioactive method for detecting specific HPV strains by hybridising HPV DNA from clinical samples with complementary RNA sequences in the kit that annealed together forming the detectable 'hybrid' molecule. Detection of HPV was achieved via antibodies that 'captured' the DNA-RNA hybrids created from an HPV-infected sample. Digene's HC test was more specific than rival PCR-based tests that were seen as too sensitive for clinical use (Poljak *et al.* 1999).

However, the HC was not simply a technically superior assay; as Digene's patent-protected proprietary platform it embodied a strategy of technological autarky. Many molecular diagnostics companies produce reagents that can be run on the platforms of their commercial competitors, but Digene chose to create its own self-contained instrumentation system to ensure they were not dependent on potential rivals: '[...] we didn't want to be subject to the vagaries of the other company who was at any moment going to be our competitor [...]' (a Digene executive).

The creation of a proprietary platform technology was a commercial advantage whatever HC was used for but, in the HPV market, Digene sought more than simply independence from rival firms through *platform* patents; it wanted to create a barrier to market entry through *biomarker* patents. HC was designed to identify ten high-risk HPV types, and Digene had patented or exclusively licensed five of those types for diagnostic use in the USA. They extended their legal monopoly with the second-generation version of HC, which contained three more HPV types, of which one was exclusively licensed to Digene.

Demonstrating clinical utility

The clinical scepticism that had greeted BRL-LT's ViraPap in 1988 was sufficient to suggest that a superior assay protected by platform and biomarker patents was no guarantee of commercial success; market acceptance would require robust evidence that demonstrated the clinical value of HPV testing in cervical cancer screening. To that end, Digene participated in multiple large head-to-head clinical studies against the Pap test during the 1990s, in collaborations with charities, government departments, universities and research institutes across the world. The clinical evidence base for their test grew incrementally, but was still limited when the first-generation Hybrid Capture test gained FDA approval in 1995. However, by 2003 Digene claimed that it had participated in studies involving 'an aggregate of approximately 90,000 women on four continents' (Digene 2002, p. 12) and the firm had increased their annual R&D expenditure to $10.6m per year. A significant portion of this sum was spent on platform development and on extending the range of tests the company offered on the platform. Nevertheless, Digene was putting some of its money into HPV clinical studies – estimated at somewhere between $20–30m in the 1990s (a former Digene executive). This level of investment was unusual; diagnostics companies have not traditionally invested significant sums to demonstrate the clinical validity and utility of new biomarkers.

Using the data collected from ISI WoS (see the Methods section), an exploratory analysis of Digene's co-authorships in scientific articles reveals how this R&D programme placed the firm at the centre of a global research network. Figure 5.2 depicts Digene's ego-network dynamics[4] across a series of three-year time windows. While Digene's ego-network clearly grows over time, the strong inter-organisational links with the National Cancer Institute, Johns Hopkins University and Kaiser Permanente are particularly evident, with repeated publications within and across time periods. Digene collaborated with the private healthcare system Kaiser Permanente and the National Cancer Institute on two studies crucial to adoption of the HC2 test for use in ASC-US triage.

As suggested earlier, this clinical research activity was distinct from the more strictly epidemiological studies that were focused on the prevalence of different HPV types and their causative association with cervical cancer. Progress in this strand of research was a prerequisite for acceptance of HPV as the necessary cause of cervical cancer in the wider scientific and clinical communities beyond the growing HPV research community. Two major milestones in this regard came mid-decade: a 1995 paper by Bosch *et al.* reported the findings of the International Biological Study on Cervical Cancer (IBSCC), a transnational study spanning 22 countries, which found that HPV was present in 93 per cent of invasive cervical cancer samples. Official recognition of their findings came when the International Agency for Research on Cancer declared HPV 16 and 18 (the two most prevalent high-risk HPV strains) to be carcinogenic agents (IARC 1995).

Mid-decade was a significant period for Digene as well. In preparation for its public flotation the company was advised by its bankers that it would not have strong appeal to investors if it presented itself as a broad-based molecular diagnostics company. Two other diagnostics companies had recently successfully floated, branding themselves as women's health companies, and Digene's advisers recommended that they ride two investment bandwagons at once - women's health and DNA - by focusing on HPV as their lead product. This external guidance resolved the firm's internal debate about the relative importance of the HPV test as its principal focus.

In 1999 Digene gained FDA approval for use of the HC2 test in ASC-US triage. At this stage the firm was taking a cautious approach to molecularisation of cervical cancer screening in the US market, seeking to support rather than supplant the Pap test. FDA approval of Hybrid Capture 2 marks the end of the development phase for Digene's technology (although earlier HPV tests for HPV testing such as BLT's had already passed the development phase but failed in the market).

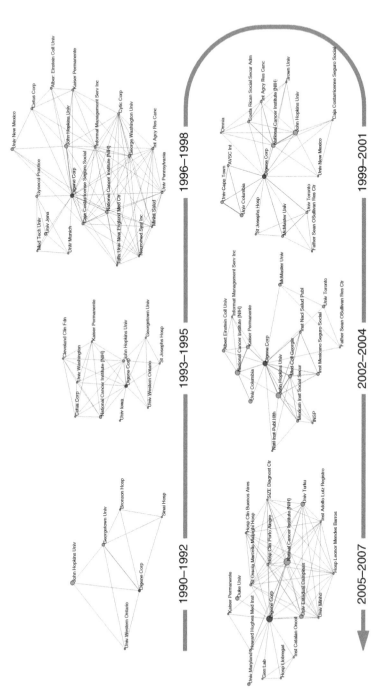

Figure 5.2 Digene's ego-network dynamics (co-authorship network). The size of nodes is proportional to the number of scientific articles related to HPV diagnostics that the given actor published in the relative time window.

Source: authors' elaboration on the basis of ISI Web of Science data and by using the Pajek software package.

FDA approval was a major milestone in the strand of HPV research focused on demonstrating the clinical utility of HPV testing, but 1999 also marked the final confirmation of the epidemiological investigation of the role of HPV in cervical cancer. Researchers in the global IBSCC study had reanalysed their HPV negative samples on the assumption that the 93 per cent prevalence rate they had reported in 1995 was lower than the true prevalence, probably as a result of inadequate specimens or HPV DNA integration. On reanalysis they found that the prevalence of HPV in their cervical cancer specimens was 99.7 per cent. They published their new findings in the *Journal of Pathology* with the bold and simple title: 'Human Papillomavirus is a necessary cause of cervical cancer worldwide' (Walboomers, *et al*. 1999). The concluding paragraph emphasised the scientific significance of the finding: 'this would indicate the highest worldwide attributable fraction ever identified for a specific cause of a major human cancer' and its clinical implications: 'our results reinforce the rationale for HPV testing in combination with, or even instead of, cytology in population-based screening programmes' (Walboomers *et al*. 1999, p. 18).

Adoption 1999–2003

Leading figures in the global HPV research community may have been making bold claims that HPV testing could replace the Pap smear but, in 1999, it was not clear whether Digene could build a sustainable market for their test in the modest role of ASC-US triage. However, by 2003 HC2 was becoming established in cervical cancer screening in the USA, Digene was nearing profitability, and commercial rivals were entering the market with their own HPV tests.

Early adoption of the first version of HC occurred even before FDA approval of HC2 in 1999, but gained a subsequent boost from new publications supporting use of the second-generation test, including a study Digene had funded with Kaiser Permanente (a large US healthcare provider) and the NCI's ALTS data. The ALTS trial was a major $21m study funded by the National Cancer Institute for which Digene provided test kits free of charge. The ALTS data was not part of the evidence submitted to FDA for approval of HC2 as the trial was still ongoing, but it is widely regarded as crucial for the subsequent endorsement of the test by professional bodies.

The commercial failure of the ViraPap had demonstrated that FDA approval was not necessarily sufficient to gain clinical adoption; overcoming clinical resistance would require backing from leading clinicians and scientists, and in the contemporary healthcare system the most effective form of professional validation is endorsement in a clinical guideline. In the last two decades the emergence of evidence-based medicine (EBM) has fuelled the growing importance of clinical practice guidelines. It has been estimated that 1,000 new guidelines are produced each year in the USA (Timmermans and Berg 2003, p. 7).

The first professional endorsement came from the American Society for Colposcopy and Cervical Pathology (ASCCP). The 2001 ASCCP guidelines stated that (providing a suitable sample was collected with the initial Pap test)

HPV testing was the preferred way of triaging women with AS-CUS, although the established options of repeat cytology or immediate colposcopy were also acceptable. The opening section of the guidelines noted that there was evidence to suggest that, in combination, the new technologies of HPV testing and liquid-based cytology 'are attractive alternatives for managing women with certain types of cytological abnormalities' (Wright *et al.* 2002, p. 2120). The guidelines presented three options: repeat cytology, colposcopy and reflex HPV testing. Reflex testing with HPV was presented as overcoming many of the disadvantages of the first two options: it did not require taking an additional specimen and would reduce the number of women referred to colposcopy by 40–60 per cent. Data from the ALTS trial is considered by many to have been critical to the guidelines:

> It [the ALTS trial] validated the performance of the test … and the recommendations based on that through ASCCP pretty much directed the growth of that test.
>
> (US LAB 2.)

The ASCCP guidelines also illustrate how Digene's network of collaborators were now playing a critical role as advocates for clinical adoption of HC2. Among the 41 members of the ASCCP working groups who contributed to the guidelines, there were seven individuals disclosing links with Digene, ranging from study grants to honoraria and consultancy work, including Tom Wright, the lead author. Similarly, five of the 38 working group members for the ACS guidelines disclosed some link to Digene.

The influence of Digene's network of collaborators is further demonstrated by an editorial, which accompanied the 2001 ASCCP guidelines, written by Mark Stoler, a senior expert in cytopathology and a participant in the ALTS trial. While the guidelines only stipulated that testing should be done 'using a sensitive molecular test', Stoler's editorial asserted that because the guidelines only drew on evidence from studies that used Digene's HC2 test, this was now the *de facto* standard for HPV testing (Stoler 2001).

Digene now had a growing clinical evidence base, externally validated through peer review, FDA approval and endorsement in clinical guidelines. To speed adoption of their test, Digene increased their investments in sales and marketing. The traditional marketing route for diagnostic companies is to enrol the support of laboratory directors who then promote new tests to physicians. However, Digene employed a dedicated sales force directly targeting physicians, a strategy seen by some investors as essential to drive rapid adoption of new molecular diagnostics (from personal communication with US venture capitalist, August 2006). This was a significant break with tradition: 'In contrast to pharmaceutical manufacturers, neither test manufacturers nor laboratory service providers generally have large, sophisticated marketing teams targeting physicians, health plans and patients' (Ramsey *et al.* 2006, p. 198).

By 2003 Digene had gained near universal insurance coverage for use of its test in ASCUS triage, acquired 62 per cent of the potential AS-CUS triage market in the USA, its customers included all the major US reference laboratories, and the company had finally become profitable. Digene's successful creation of a market for HPV testing in the USA piqued the interest of its commercial rivals. Using the technology it had acquired from Oncor, Ventana launched an IHC test in 2001 (without FDA approval). Senior management at Roche began to pay more attention to HPV testing around 1999, their interest sparked by the clinical data from the Kaiser Permanente study and the 1999 Walboomers paper that had confirmed the role of HPV as a necessary cause of cervical cancer.

However, HPV testing was not without its critics. Senior experts had criticised the early ALTS data (Herbst *et al.* 2001), and in 2003 the United States Preventive Services Task Force issued a new guideline stating that there was insufficient evidence to support use of HPV testing. Even in Kaiser Permanente, which had been a rapid early adopter of HC2, there were sceptics. Kaiser's HPV advocates were based in its Northern California division but, in Southern California, Neal Lonky and colleagues published a study of HPV-based ASC-US triage in 2003, concluding that the sensitivity of HC2 was a cause for concern:

> Although less complicated than colposcopy, the Hybrid Capture II triage algorithm for ASCUS will under-diagnose some women with high-grade CIN, when compared with colposcopy.
>
> (Lonky *et al.* 2003.)

The final critical development to note in this period was not a molecular HPV test but a refinement of the Pap smear: Liquid Based Cytology (LBC). The leading manufacturer of LBC kits was a US firm called Cytyc, which had first gained FDA approval for their ThinPrep test in 1996 (the same year in which the ALTS trial began). Its chief commercial rival AutoCyte had its SurePath test approved by FDA in 1999. Clinical adoption of LBC grew rapidly following the publication in 1999 of a study of the new technology carried out by Quest, one of the two major US reference laboratories. FDA approval of the Digene test was based on sample collection using either Digene's proprietary collection kit and preservation medium or using Cytyc's technology (that, to date, the FDA has not approved). The two firms had collaborated on the ALTS trial and in 2001 Digene and Cytyc announced that they would embark on a joint promotion campaign for their respective products (Anonymous 2001). Cytyc even planned to acquire Digene, but the Federal Trade Commission blocked the transaction.

Growth (2003–2014)

Following adoption of HC2 for triage, Digene focused on the larger screening market. Since 2003, growing sales, the market entry of other companies, and endorsement in new clinical guidelines demonstrate widespread validation of HPV testing. In 2014 FDA approval of Roche's COBAS HPV test for use as the sole primary screening test, has created the possibility that cytology may lose its primacy in cervical cancer screening.

Digene's strategy to expand its market

Use of Digene's test for triage of AS-CUS cases was the low-hanging fruit of HPV testing. It exploited a chief clinical weakness of cytology, namely the large number of ambiguous results requiring further follow-up, but did not challenge cytology's status as the gold standard for cervical cancer screening. ASCUS triage was, moreover, a relatively small market. Perhaps unsurprisingly, most of Digene's R&D investment was focused on the more lucrative primary screening market, funding studies where HPV testing was a routine adjunctive screen alongside cytology or an alternative to it. The company's intent was made clear in an article in the *Washington Post* in January 2001 in which Evan Jones, Digene's CEO and chairman, stated: 'Our goal is to replace the Pap smear' (Chea 2001). However, at this stage there was scepticism even among leading figures in Digene's network of collaborators; the same news report quoted the NCI's Mark Schiffman expressing caution about a shift away from cytology: 'The Pap smear has a long tradition of reducing cervical cancer... We don't want to dislodge good cervical cancer screening unnecessarily'. Adjunct screening, also known as co-testing, was a compromise between those advocating the use of HPV testing as the sole primary screening test and influential figures like Schiffman who were more cautious.

In 2002 guidelines from the American Cancer Society (ACS) recommended HPV testing as an adjunctive screen in women over 30, and in 2003 this indication gained FDA approval and clinical endorsement in a guideline from the American College of Obstetricians and Gynecologists (ACOG 2003). The adjunct screening version of the Hybrid Capture test was branded as DNAwith Pap. By 2003 the company had increased its annual HPV testing revenue 64 per cent to $40m and it estimated that the approval for co-testing had created a potential market for its product of $400m. Its 2003 annual report stated that it had achieved insurance coverage for more than 50 million lives for co-testing.

Long-time Digene collaborators Kaiser Permanente were early adopters, again demonstrating the importance of key inter-organisational links (such as the research collaborations indicated in Figure 5.1), but also illustrating the significance of institutional structures in building early adoption:

> The Kaiser Permanente HMO model was ideally suited in some ways to making such a drastic change. First, because Kaiser directly cares for its paying members, it did not have to convince outside payors to cover the additional test. Second Kaiser employs its physicians, so if the administration wanted to add a new laboratory test, it could do so without formal buy-in, though implementing the test requires the understanding and cooperation of the ob/ gyn providers.

> (Southwick 2004.)

Even in the organisational structure of Kaiser Permanente, implementation of this new testing protocol met with initial clinical resistance (despite an educational programme to persuade clinicians of its virtues), and elsewhere clinician adoption was slow. The advantage claimed for the combined screening technologies was that women with negative results need not be tested again for three years (in the USA annual screening is routine). But lengthened intervals was a significant shift in practice:

> For many women, the only reason they see a clinician is to have a Papanicolaou smear taken. Although that specimen might not be necessary, many other events, critical to the maintenance of health, occur during the Papanicolaou smear visit. These include the measurement of blood pressure and weight, review of vaccinations and medications, and counselling regarding a healthy lifestyle.
>
> (Noller *et al.* 2003.)

Digene developed an multi-pronged marketing strategy for the commercial launch of DNAwithPap, including: collaborating with government agencies, professional bodies and women's advocacy groups to communicate their support for the test among professionals and the public; partnering with laboratories to co-market the test to physicians and payors; and creating demand through education programmes driven by their physician detailing organisation and independent third-party organisations to educate physicians and women about HPV testing (Digene 2003, pp. 14–15).

Digene funded Women in Government (WiG), a Washington DC-based non-governmental organisation, who launched the Challenge to Eliminate Cervical Cancer Campaign in 2004. Women in Government targeted state policymakers with the message that cervical cancer was preventable and that 'all women [should] have access to the most advanced and appropriate cervical cancer prevention technologies, regardless of their socioeconomic status'. By 2006 California, Maryland, North Carolina, New Mexico, Texas and Virginia had amended their legislation to mandate coverage of HPV testing. Other NGOs working with Digene included the Coalition of Labour Union Women, Hadassah (the Women's Zionist Organisation of America) and the Women's National Basketball Federation. Digene even had a musical campaign, the Pop Smear tours organised by Yellow Umbrella and led by Christine Baze, a musician and cervical cancer survivor.

However, despite a promising level of early adoption in the first 18 months after FDA approval, growth began to stall thereafter. Digene responded by adopting a new marketing strategy: direct-to-consumer advertising. The company reported that it planned to spend between $3–5m in 2005 on adverts on television and in women's magazines, which carried stark messages such as 'You're not failing your Pap test, but it might be failing you' (Grebow 2005). A report in the *Washington Post* highlighted professional disquiet with the campaign; Marcia Angell, a former editor of the *NEJM*, suggested that the ads '…were just trying to frighten women' and questioned the legitimacy of DTC advertising: 'It's marketing, and it creates demands that should just be between patients and their doctors' (Rosenwald 2005). Digene defended its advertising campaign as a legitimate educational tool that raised women's awareness of the cause of cervical cancer. A consumer survey conducted six months after launch of the campaign found that '82 per cent of women exposed to the ads said they had talked or planned to talk with their doctors about the HPV test ... [and that] 51 per cent of physicians surveyed said they had seen a significant increase in requests for the HPV test'. Perhaps, most tellingly, while test revenues increased 42 per cent over the six-month period in the USA as a whole, growth was 85 to 115 per cent in the metro-politan areas, which had been targets for both print and TV ads (Luchtefeld 2006).

A molecular monopoly - protecting market share

Despite continuing clinical resistance the HPV market has grown. Most insurance companies cover HPV testing; it is mandated under the Affordable Care Act for women 30 or over as a service that insurers must provide without charging women copayment fees, and is available through the public Medicaid system in most states. In 2008 industry estimates suggested that more than 10 million tests were being performed annually and that the market had grown 40 per cent in each of the past five years (Fischer 2008). Yet at the end of the fiscal year for 2006, Digene estimated their penetration of the total potential US HPV testing market to be approximately 18 per cent. In 2007 an industry news report quoted Roche Molecular as stating that the co-testing market had reached 20 per cent (Bruderlin-Nelson 2007). By 2009 Qiagen (who had acquired Digene) were stating that the screening market had reached 30 per cent penetration (Clancy 2009). The figures from the laboratory at Johns Hopkins University illustrate how much more slowly the contesting market grew compared with ASCUS triage.

Furthermore, despite being the only FDA approved test until 2009, not all HPV testing was performed using Digene's HC2. Some pathology labs were develop-ing their own tests from component reagents sold by other firms The FDA had long permitted diagnostics companies some latitude, permitting them to sell what it termed Analyte Specific Reagents (ASRs) to laboratories without regulatory approval, providing the firm made no marketing claims and gave no instructions for use. Responsibility for validation of tests developed using ASRs thus falls on the laboratories, rather than the diagnostics firms, and laboratories must report test results with a disclaimer noting that the test has not been FDA-approved. By 2004 the ASR rule was being exploited by a number of firms marketing HPV ASRs including Roche, Third Wave and Ventana. However, in 2004, both Ventana and Roche received communications from the FDA stating that their HPV ASRs would require pre-market authorisation as Class III devices. Roche removed their products from the market and sought FDA approval, but Ventana was permitted to keep its ASRs on the market subject to changes to its product literature (Anonymous 2004). A 2006 survey suggested that other tests were being used in 19.1 per cent of US labs, either alongside Digene's HC2 test, or instead of it (Moriarty *et al.* 2008).

Regulatory approval was only a partial barrier to market entry but Digene also benefited from interventions by their collaborators to limit the spread of rival

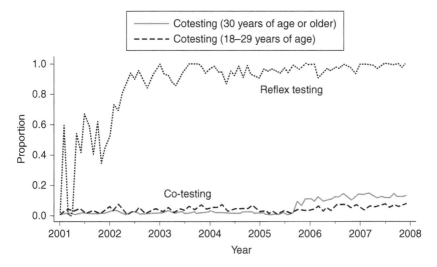

Figure 5.3 Monthly proportions of human papillomavirus (HPV) reflex testing and cotesting with Pap for Pap specimens processed by the Johns Hopkins Hospital Division of Cytopathology, 2001–2007.

Source: Phelan, 'Trends of human papillomavirus testing'. *Obstet Gynecol* 2011.

technologies lacking FDA approval. Controversy about such tests was raised in 2005 in an article in *CAP Today*, the magazine of the College of American Pathologists, which quoted Digene's NCI collaborator, Marc Schiffman:

> I do not want to see decades of careful research lessened in their impact by sloppy application or sloppy thinking. If a well-meaning laboratory applies an HPV test that doesn't work right, then a beneficial technology has just been made malignant.

Also commenting was Atilla Lorincz, Digene's Chief Scientific Officer: 'We spent tens of millions of dollars validating this test. For someone to come along and run 70 or 80 patients verges on the insult to everybody'.

This position was reinforced by the endorsement of HC2 in clinical guidelines. The 2002 ACS guidelines that recommended co-testing stated that HC2 was the only test approved by FDA and that only evidence from studies using HC2 had been used in the development of the guidelines. In 2006 new ASCCP guidelines went further, suggesting less validated tests 'may increase the potential for patient harm' (Wright *et al.* 2007, p. 347). These statements reinforced the status of Hybrid Capture as the gold standard for HPV testing. Thus two forms of gate-keeping – statutory licensing and professional self-regulation – were mutually reinforcing: the guidelines affirmed the importance of FDA approval, and the status of Digene's

HC2 test as the only FDA-approved test legitimated the focus on the company's proprietary technology within the guidelines. The guideline authors stressed the need for HPV tests to be thoroughly validated and this was an implicit endorsement of FDA's position that all HPV tests should be treated as Class III, high-risk devices subject to the most rigorous PMA approval process.

Digene also sought to restrict competition by using intellectual property rights as a barrier to market entry. We have already referred to Digene's legal monopoly on key high-risk HPV strains. Between 2001 and 2009 this monopoly was defended in a series of US law suits with rivals Beckman Coulter, Gen-Probe, Roche and Third Wave.

Rivals gain FDA approval

The high bar of FDA approval and the power of Digene's patent position combined to exclude any serious competition in the US market for a decade. Roche bought the Institut Pasteur's patents (including patents on HPV types exclusively licensed to Digene) in 2002 and launched an HPV test in Europe in 2003, but the firm's attempts to gain FDA approval for its Amplicor and Linear Array HPV tests failed. Digene lacked an FDA-approved commercial rival until Third Wave gained approval for their Cervista test in 2009. Competition intensified in 2011 when the FDA approved HPV tests from Roche and Gen-Probe, the two largest molecular diagnostics companies in the USA.

Market entry by rivals to Digene was not the only signal that HPV testing was now an established market with strong growth potential - the field has also seen multiple acquisitions. In 2007 Digene was bought for $1.6bn by the German firm Qiagen, a price that demonstrated the commercial success of Digene's strategy and the perceived value of the HPV test, as Digene had little else in its development pipeline (Baker 2006). Hologic, a US firm with ambitions to be the leader in women's health diagnostics, has acquired both Third Wave (in 2008) and Gen-Probe (in 2012), giving it multiple HPV technologies. Hologic's acquisition of Cytyc (in 2007) has given the firm the combination of liquid-based cytology and HPV technologies, which Cytyc had sought to achieve when it tried to acquire Digene in 2002.

New guidelines converge on co-testing

The prospects for HPV testing were given a boost in 2012 with the publication of three new guidelines endorsing routine co-testing. Reflecting the new market reality of multiple FDA-approved tests, these guidelines no longer rely solely on evidence from studies using Digene's HC2 test, although they state that in the USA only FDA-approved tests should be used in clinical practice and recommend against the use of laboratory-developed tests.

These new guidelines drew on new evidence from four large randomised-control trials that had been published in the preceding five years. The US Preventive Services Task Force (USPTF) guideline summarised their evaluation of the evidence thus: 'Modelling studies support similar benefits of co-testing every 5 years and cytology every 3 years, demonstrating small differences in expected cancer cases (7.44 *vs.* 8.50 cases, respectively) and cancer deaths (1.35 *vs.*1.55 deaths, respectively)' (Moyer 2012, p. 887).

Based on this new data the USPTF (which had previously rejected any use of HPV testing) now supports co-testing, although it does not say it is the preferred option (Moyer 2012). By contrast a joint guideline from the American Cancer Society, the American Society for Colposcopy and Cervical Pathology, and the American Society for Clinical Pathology (Saslow *et al.* 2012) and a new guideline from the American College of Obstetricians and Gynaecologists (ACOG 2012) went further, both stating that for women 30 or over, co-testing is preferable to the use of cytology alone (although the preference for co-testing was classed as a 'weak' recommendation in the appendix to the ACS guidelines).

The three guidelines shared a stated aim: to reduce unnecessary screening. A common recommendation was no cervical cancer screening in women under 21, a change which could save at least $500m a year (Morioka-Douglas and Hillard 2013). For older women, the guidelines all recommended a move away from annual screening, an extension of prior attempts to lengthen screening intervals in the USA.

However, while guidelines have helped to drive clinical adoption of HPV testing, the impact of their recommendations has been uneven. We have already seen that a proportion of laboratories did not follow the recommendation to use only FDA-approved tests, and that adoption of co-testing has been slower than for ASCUS triage. Lack of physician compliance has been investigated in a number of studies. A survey of ACOG members carried out in 2011–12 found that around half were using co-testing but that most were continuing to offer annual tests: 'Physicians felt that patients were uncomfortable with extended screening intervals and were concerned that patients would not come for annual exams without concurrent Paps' (Perkins *et al.* 2013). A 2006 survey of primary care physicians found that those who would recommend co-testing were less likely to comply with guidance on lengthened screening intervals (Saraiya *et al.* 2010), suggesting that co-testing appeals to a 'more is better' mentality. In an interview Alan Waxman, one of the authors of the 2012 guidelines, discussed one of the concerns that influence physicians: 'Healthcare providers are influenced by dramatic cases they see. This is what we call the "N of 1." You see 1 bad case and you want to direct your management for all patients to fit that case' (Terry *et al.* 2013).

Our interviewees frequently suggested that persistence of annual testing is another example of the practice of defensive medicine in the USA. However, they also alluded to a number of other factors slowing adoption of co-testing and lengthened screening intervals, the chief one being feared loss of income and another being a reluctance to have to counsel patients on a positive HPV test.

Even those involved in writing the guidelines suggested that compliance was not a straightforward matter because the new system was far more complex than the established management protocols.

the old message … Get your Pap, easy message. What's the message now? Get your pap and if abnormal, now it's get your HPV test and if you're positive, if you're negative you get into the controversy of how long to wait until doing it again, if you're positive you get into the issue of where did I get it blah-blah-blah and then if you're in because of co-testing in the US, if you have a discrepant result: HPV positive/pap negative, HPV negative/ASCUS, HPV negative/LSIL, there's all these combinations that dominate a book this size, this is a guidelines book I mean it's just too complicated ….

(USA 2.)

Beyond cytology?

Notwithstanding these unresolved issues, the ACS and USPTF guidelines also point to emerging evidence that may support the use of HPV testing as the sole primary screening test. In April 2014 Roche became the first firm to win FDA approval for this indication. The experience of co-testing would suggest that such a dramatic shift in clinical practice may be extremely difficult to achieve. Although the experts on the FDA advisory committee that reviewed the Roche submission in March 2014 were unanimous in their support for approval of primary screening in women 25 or over, there was resistance from some groups, including a coalition of 17 NGOs led by the Cancer Prevention and Treatment Fund, who wrote a joint letter to Margaret Hamburg, the FDA Commissioner, arguing that the superiority of HPV primary screening is unproven: 'It replaces a safe and effective well-established screening tool and regimen that has prevented cervical cancer successfully in the U.S. with a new tool and regimen not proven to work in a large U.S. population' (Patient, Consumer and Public Health Coalition 2015).

David Chelmow, a spokesman for ACOG, expressed several reservations at the advisory committee meeting: that the new option would further confuse clinicians still struggling to come to terms with the complex management options available under the 2012 guidelines; that there was a lack of data on some of the issues which a transition to primary screening would raise, such as what to do with women under 25; and finally, the absence of data on the comparative benefits of co-testing *vs.* primary HPV screening.

Following the approval announcement, ASCCP indicated that it was working with other professional bodies on an interim guidance to address the new option (ASCCP 2014), but at the September 2014 meeting of the International Papillomavirus Society it was revealed that progress towards a new guidance is slow, with experts in disagreement on a number of issues, including the age at which to start HPV primary screening, the appropriate screening interval, and the optimal triage protocol for HPV positive women.

Conclusions

HPV testing has been available for 25 years, but it has yet to supplant cytology; instead, the molecularisation of cervical cancer screening has required an accommodation with the established socio-technical regime. This reflects the wider picture in oncology, where diagnosis still relies on morphological examination of tumour biopsies, despite a growing number of molecular diagnostics. This evolutionary model would appear to be consistent with the broader history of diagnostic innovation in the twentieth century: 'The newer modes of analysis have not necessarily replaced the older ones. In many ways, the history of these techniques has been one of continuous accretion …' (Amsterdamska and Hiddinga 2003, p. 426). This mirrors the uptake of biotechnology in pharmaceutical R&D, where biotechnological techniques have reinvigorated traditional small molecule drug discovery, as well as supporting the development of novel biologic therapies (Hopkins *et al.* 2007). As the historian David Edgerton has suggested, we should not mistake the first appearance of a technology for its triumphant dominance, and older technologies may be increasing in use, even as new ones are growing in popularity (Edgerton 2006, p. 31–2).

Notwithstanding these continuities, the history of HPV testing in the USA suggests that the molecularisation of cancer diagnosis is accompanied by some important changes in the diagnostic innovation process. The emergence of the Pap smear exemplified two key characteristics of diagnostic innovation in the twentieth century – publicly funded discovery and development of a new biomarker, and the creation of a new sub-specialty within the pathology profession. The Pap smear emerged from basic research conducted in an academic setting. Its potential clinical application was first explored in the 1920s during a period when pathologists were gradually wresting control of cancer diagnosis from the surgical profession (Lowy 2010, p. 20), and its development exemplified a model of diagnostic innovation in which the public sector played the primary role.

The history of laboratory diagnostics in the twentieth century was one of professionalisation and the creation of new sub-disciplines such as microbiology and radiology (Amsterdamska and Hiddinga 2003). The Pap smear exemplified that trend, predicated as it was on the creation of a new cadre of cytology specialists (Casper and Clarke 1998). The promotion of Pap testing was largely carried out by non-profit organisations such as the American Cancer Society (ibid.). By contrast, HPV testing exemplifies a new trend: the increasing importance of diagnostic companies in the development and diffusion of innovative molecular diagnostics.

The development of Hybrid Capture involved a dynamic process of deepening engagement between Digene and a variety of actors in the public sector. In the first place it suited clinical researchers to collaborate with Digene, who might either subsidise or pay for clinical trials. Secondly, as early data demonstrated the robustness of their test, it gained credibility as a dependable research tool whose use in multiple trials across the globe could be expected to produce reliable standardised data that could be subject to cross-comparison and meta-analysis. Finally, as the research community began to produce findings that indicated a possible role for HPV testing in cervical cancer screening, these researchers became advocates for the clinical use of the HC2 technology as the only HPV test that had both proven its value in multiple large clinical trials and carried the imprimatur of FDA approval. Digene thus harnessed a growing interest in the clinical potential of HPV testing to create an international research network focused on demonstrating the clinical utility of their proprietary technology.

The establishment of Hybrid Capture as the tool of choice for detecting the presence of HPV was not simply a question of which company had the superior technology but was also a matter of intellectual property rights. Digene's success was achieved not only by enrolling key supporters to drive clinical adoption, but also by (at least temporarily) excluding competitors from joining the network.

This case illustrates how a young, relatively small diagnostic company can become the orchestrator of global networks involving research scientists, funding agencies, laboratory directors, clinicians, patients and regulatory agencies. This is a new 'systems integration' role for diagnostics companies, mirroring that seen in pharmaceutical firms (Hopkins *et al.* 2007), illustrating how molecularisation involves not simply a greater role for industry but also a shift in business models. Crucial aspects of Digene's commercial strategy - patenting biomarkers to try and gain a period of market exclusivity, marketing to physicians, consumer advertising, investing heavily in studies to demonstrate the clinical utility of a test – were all relatively novel to the IVD industry, and some at least were not without controversy.

However, the novelty of the commercial drivers should not deflect attention from the continuities with the Pap story. Public bodies played a pivotal role in the promotion of Pap testing and were also central to the clinical adoption of the HPV test, especially the NCI who funded the ALTS trial and (alongside the ACS) then championed Digene's HC2 technology as the only robustly-validated HPV test. Collaboration with industry has thus reinforced the authority of the established network of public sector actors in US cancer screening.

Acknowledgement

This chapter contains excerpts that have been reprinted from Hogarth, S., Hopkins, M. and Rodriguez, V. (2012) 'A molecular monopoly? HPV testing, the Pap smear and the molecularisation of cervical cancer screening in the USA'. *Sociology of Health and Illness*, 34(2), pp. 234–50. Copyright 2012. with permission from John Wiley and Sons.

Notes

1 We limited the search of the keywords and their combinations to those in the titles of scientific articles. We specifically used the following search string in the WoS: (TI=HPV* or TI="Human Papilloma Virus*" or TI="Human Papillomavirus*" or TI="Human Papilloma*virus*") and (TI=Cervical or TI=Cervix) and (TI=diagnos* or TI=test* or TI=assay or TI=detect* or TI=screen* or TI=predict*). It is worth noting that while it is also possible to search the keywords in the abstracts of scientific articles, which would allow more records to be retrieved, this would increase the number of records not closely related to the HPV diagnostics (Rotolo *et al.* 2014).

2 The cleaning of organisations' names was performed by using The Vantage Point software.
3 The study required a full 'patent landscape' to reveal commercially active organisations - see Hoschar *et al.* (2007) for a description of patent search methods. Some of these results are discussed elsewhere in Hogarth *et al.* (2012) but are not reproduced here.
4 An ego-network (or ego-centred network) 'consists of a focal actor, termed ego, as set of alters who have ties to ego, and measurements of the ties among these alters' (Wassermann and Faust 1994, p. 42).

References

ACOG (2003) 'Cervical cytology screening'. *ACOG Practice Bulletin*, 45, August
ACOG (2012) Screening for cervical cancer. Practice Bulletin No. 131. American College of Obstetricians and Gynecologists. *ObstetGynecol* 120, pp. 1222–38
Amsterdamska, O. and Hiddinga, A. (2003) 'The analyzed body' in R. Cooter and J. Pickstone (Eds) *Companion to Medicine in the Twentieth Century*. London: Routledge, pp. 417–434
Anonymous (2001) 'Cytyc and Digene announce exclusive co-promotion agreement'. *Business Wire*, 18 Jan.
Anonymous (2004) 'Ventana's HPV test retains ASR status'. *IVD Technology*, 1 June. Available online at: www.ivdtechnology.com/article/ventana per centE2 per cent80 per cent99s-hpv-test-retains-asr-status
The American Society for Colposcopy and Cervical Pathology (ASCCP) (2014) 'FDA announcement expands options for cervical cancer screening'. Maryland: ASCCP
Baker, M. (2006) 'New-wave diagnostics'. *Nature Biotechnology*, 24, 8, pp. 931–938
Blume, S. (1992) *Insight and Industry: On the Dynamics of Technological Change in Medicine*. Cambridge, MA: MIT Press
Bosch, F. X., Manos, M. M., Muñoz, N., Sherman, M., Jansen, A. M., Peto, J., Schiffman, M. H., Moreno, V., Kurman, R. and Shah, K. V. (1995) 'Prevalence of human papillomavirus in cervical cancer: a worldwide perspective'. *J Natl Cancer Inst*, 87: pp. 796–802
Bruderlin-Nelson, C. (2007) 'HPV test market expected to boom'. *IVD Technology*, May
Casper, M. and Clarke, E. (1998) 'Making the Pap smear into the "right tool" for the job: cervical cancer screening in the USA, circa 1940–1995'. *Social Studies of Science*, 28, pp. 255–290
Chea, T. (2001) 'New cervical cancer test gains backers'. *Washington Post*, 31 January
Clancy, S. (2009) 'Inside track'. *Clinical Lab Products*, April
Corliss, J. (1990) 'Utility of ViraPap remains to be established'. *Journal of National Cancer Institute* 82(4), pp. 252–3
Cox, T. and Cuzick, J. (2006) 'HPV DNA testing in cervical cancer screening: from evidence to policies.' *Gynaecologic Oncology*, 103, pp. 8–11
Dehn, D., Torkko, K. C. and Shroyer, K. R. (2007) 'Human papillomavirus testing and molecular markers of cervical dysplasia and carcinoma'. *Cancer*, 111, 1, pp. 1–14
Digene (1996) 'S-1/A, S-1 Amendment # 3'. 21 May
Digene (2002) *Annual Report*. Available online at: http://secwatch.com/dige/10k/annual-report/2002/9/30/4393963
Digene (2003) *Annual Report*. Available online at: http://web.archive.org/web/20040723122952/http://www.digene.com/corporate01/Annual_Report/digene AR03.pdf

Edgerton, D. (2006) *The Shock of the Old: Technology and Global History since 1900*. London: Profile Books.

Fischer, A. H. (2008) 'Practice patterns in HPV testing'. *CAP Today*, August

Gardner, K. (2006) *Early Detection: Women, Cancer and Awareness Campaigns in the Twentieth-century United States*. Chapel Hill: University of North Carolina Press

Gillis, J. (1998) 'Oncor's biotech woes offer a cautionary tale'. *Washington Post*, 9 November

Grebow, M. (2005) 'Digene ads stoke interest in direct-to-consumer marketing'. *IVD Technology*, June

Herbst, A. L., Pickett, K. E., Follen, M. and Noller, K. L. (2001) 'The management of ASCUS cervical cytologic abnormalities and HPV testing: a cautionary note'. *Obstet. Gynecol.* 98(5), pp. 849–851

Hogarth, S., Hopkins, M. and Rodriguez, V. (2012) 'A molecular monopoly? HPV testing, the Pap smear and the molecularisation of cervical cancer screening in the USA'. *Sociology of Health and Illness*, 34(2), pp. 234–50

Hopkins, M., Martin, P., Nightingale, P., Kraft, A. and Mahdi, S. (2007) 'The myth of the biotech revolution: an assessment of technological, clinical and organisational change'. *Research Policy*, 36, 4, pp. 566–589

Hoschar, A. P., Tubbs, R. R. and Hunt, J. L. (2007) 'Clinical testing for low and high risk HPV by ISH in respiratory papillornatosis'. *Laboratory Investigation*, Vol. 87 (March), pp. 223A–223A

Hughes, J. (1997) 'Whigs, prigs, and politics: problems in the historiography of contemporary science' in T. Soderqvest (Ed.) *The Historiography of Science and Technology*. Amsterdam: Harwood, pp.19–37

IARC (1995) *Monographs on the Evaluation of Carcinogenic Risks to Humans: Human Papillomaviruses*, Vol. 64. Lyon: International Agency for Research on Cancer

IARC (2012) *GLOBOCAN 2012: Estimated Cancer Incidence, Mortality and Prevalence Worldwide in 2012*. Available online at: http://globocan.iarc.fr/Pages/fact_sheets_cancer.aspx

Keating, P. and Cambrosio, A. (2003) *Biomedical Platforms: Realigning the Normal and the Pathological in Late-Twentieth-Century Medicine*. Cambridge MA: MIT Press

Lonky, N., Felix, J. MD, Yathi, M., Naidu, Y. and Wolde-Tsadik, G. (2003) 'Triage of atypical squamous cells of undetermined significance with Hybrid Capture II: colposcopy and histologic human papillomavirus correlation'. *Obstetrics and Gynaecology*, 101(3), pp. 481–489

Lowy, I. (2010) *Preventive Strikes: Women, Pre-cancer and Prophylactic Surgery*. Baltimore: Johns Hopkins University Press

Lowy, I. (2011) *A Woman's Disease: The History of Cervical Cancer*. Oxford: OUP

Luchtefeld, L. (2006) 'Driving awareness'. *MDDI Online*, September. Available online at: www.mddionline.com/article/driving-awareness

Manos, M. M., Wright, D. K., Ting, Y., Broker, T. R. and Wolinsky, S. M. (1989) *US patent no PCT/US1989/003747*. Filed 1989, patent published 1990

Moriarty, A. T., Schwartz, M. R., Eversole, G., Means, M., Clayton, A., Souers, R., Fatheree, L., Tench, W. D., Henry, M. and Wilbur, D. C. (2008) 'Human papillomavirus testing and reporting rates: practices of participants in CAP's interlaboratory comparison program in gynecologic cytology in 2006'. *Arch Pathol Lab Med*, 132, pp. 1290–1294

Morioka-Douglas, N. and Hillard, P. (2013) 'No Papanicolaou tests in women younger than 21 years or after hysterectomy for benign disease'. *JAMA Internal Medicine*, online April. doi:10.1001/jamainternmed.2013.316

Moyer, V. A. (U.S. Preventive Services Task Force) (2012) 'Screening for cervical cancer: U.S. Preventive Services Task Force recommendation statement'. *Ann Intern Med.*, 156: pp. 880–91, W312

Noller, K. L., Bettes, B., Zinberg, S. and Schulkin, J. (2003) 'Cervical cytology screening practices among obstetrician–gynaecologists'. *Obstetrics and Gynaecology* 102(2), pp. 259–265

Oncor (2006) *Annual report, Form 10-K.* Available online at: http://web.archive.org/liveweb/http://www.sec.gov/Archives/edgar/data/806637/0000806637-97-000001.txt

Patient, Consumer, and Public Health Coalition (2015) 'Coalition letter to FDA Commissioner about approving Cobas HPV test alone (without Pap smear)'. Available online at: www.stopcancerfund.org/policy/testimony-briefings/coalition-letter-hpv-test

Perkins, R. B., Anderson, B. L., Gorin, S. S. and Schulkin, J. A. (2013) 'Challenges in cervical cancer prevention: a survey of U.S. obstetrician-gynecologists'. *American Journal of Preventive Medicine* 45(2), pp. 175–181

Poljak, M., Brencic, A., Seme, K., Vince, A. and Marin, I. J. (1999) 'Comparative evaluation of first- and second-generation Digene hybrid capture assays for detection of human papillomaviruses associated with high or intermediate risk for cervical cancer'. *J Clin Microbiol.* Mar; 37(3), pp. 796–7

Ramsey, S. D., Veenstra, D. L., Garrison, L. P. Jr, Carlson, R., Billings, P., Carlson, J. and Sullivan, S. D. (2006) 'Towards evidence-based assessment for coverage and reimbursement of laboratory-based diagnostic and genetic tests'. *American Journal of Managed Care* 12(4), pp. 197–202

Reynolds, L. A. and Tansey, E. M. (Eds) (2009) 'History of cervical cancer and the role of the Human Papillomavirus, 1960–2000'. *Wellcome Trust Witness Seminar Series*, 38

Rosenwald, M. (2005) 'Digene's ads take their case to women'. *Washington Post*, Monday, 21 March

Rotolo, D., Rafols, I., Hopkins, M. and Leydesdorff, L. (2014) 'Scientometric mappings as strategic intelligence for tentative governance of emerging science and technologies'. *SPRU Working Paper Series*, 2014–10, pp. 1–40. Available online at: www.sussex.ac.uk/spru/swps2014-10

Saraiya, M., Berkowitz, Z., Yabroff, K. R., Wideroff, L., Kobrin, S. and Benard, V. (2010) 'Cervical cancer screening with both Human Papillomavirus and Papanicolaou testing *vs* Papanicolaou testing alone: what screening intervals are physicians recommending?' *Arch Intern Med.* 170 (11), pp. 977–986

Saslow, D., Solomon, D., Lawson, H. W., Killackey, M., Kulasingam, S. L., Cain, J., Garcia, F. A., Moriarty, A. T., Waxman, A. G., Wilbur, D. C., Wentzensen, N., Downs, L. S. Jr, Spitzer, M., Moscicki, A. B., Franco, E. L., Stoler, M. H., Schiffman, M., Castle, P. E. and Myers, E. R. (2012) 'American Cancer Society, American Society for Colposcopy and Cervical Pathology, and American Society for Clinical Pathology screening guidelines for the prevention and early detection of cervical cancer'. *Am J ClinPathol* 2012, 137, pp. 516–42

Schiffman, M., Wentzensen, N., Wacholder, S., Kinney, W., Gage, J. and Castle, P. (2011) 'Human papillomavirus testing in the prevention of cervical cancer'. *JNCI*, 103(5), pp. 368–383

Shibata, D., Arnheim, N. and Martin, W. J. (1988) 'Detection of human papilloma virus in paraffin-embedded tissue using the polymerase chain reaction'. *Jnl Exp Med.* 167(1), pp. 225–30

Southwick, K. (2004) 'Kaiser roll: the push for more HPV screening'. *CAP Today*, July

Stoler, M. (2001) 'HPV for cervical cancer screening: is the era of the molecular Pap smear upon us?' *Journal of Histochemistry and Cytochemistry*, 49, 9, pp. 1197–1198

Terry, N., Darragh, T. M. and Waxman, A. G. (2013) 'New ASCCP Guidelines: equal management for equal risk'. *Medscape*. Available online at: www.medscape.com/view-article/782362_6 (last accessed 28 May 2013)

Timmermans, S. and Berg, M. (2003) *The Gold Standard*. Philadelphia: Temple University Press

Titus, K. (2005) 'Making a valid point about HPV tests'. *CAP Today*, September

Titus, K. (2006) 'How Pap test automation is playing out'. *CAP Today*, October

Van de Ven, A. H. (2007) *Engaged Scholarship*. Oxford, UK: Oxford University Press

Walboomers, J. M. M., Jacobs, M. V., Manos, M. M., Bosch, F. X., Kummer, J. A., Shah, K. V., Snijders, P. J. F., Peto, J., Meijer, C. J. L. M. and Muñoz, N. (1999) 'Human papillomavirus is a necessary cause of invasive cervical cancer worldwide'. *J Pathol*. Sep, 189(1), pp. 12–9

Wassermann, S. and Faust, K. (1994) *Social Network Analysis: Methods and Applications*. Cambridge, UK: Cambridge University Press

Wright, T. C. Jr, Cox, J. T., Massad, L. S., Twiggs, L. B. and Wilkinson, E. J. (2002) '2001 consensus guidelines for the management of women with cervical cytological abnormalities'. *JAMA*, 287, 16, pp. 2120–2129

Wright, T. C. Jr, Massad, L. S., Dunton, C. J., Spitzer, M., Wilkinson, E. J. and Solomon, D. (2007) '2006 Consensus guidelines for the management of women with abnormal cervical cancer screening tests'. *American Journal of Obstetrics & Gynaecology*, October, pp. 346–355

Yuxiang, G., Jay, G., Khorshidi Hossein, S., Mishra Nrusingha, C. and Szweda, P. for Oncor Inc. (1996) *Modified nucleotides for nucleic acid labelling*. Patent Number WO 1996041006 A1 (APPLICANT NAME: ONCOR INC)

zur Hausen, H. (1987) 'Papillomaviruses in human cancer'. *Applied Pathology*, 5, pp. 19–24

zur Hausen, H. (2006) 'Perspectives of contemporary papillomavirus research'. *Vaccine* 24S3 (2006), pp. S3/iii–S3/iv

6   Poliomyelitis vaccine innovation

Ohid Yaqub

Introduction

The core issue addressed in this chapter was originally highlighted by Rosenberg in his 1974 critique of Schmookler's 1966 argument that innovation was induced by demand. Nearly 40 years later, his words still seem prescient to me, as more and more policy advisors rush to claim that they can help 'pull', 'procure', 'incentivise' and strengthen demand for innovation. Rosenberg acknowledged that plenty of instances exist where technical understanding was improved in response to societal demand, but he also highlighted abundant evidence where weak scientific understanding constrained innovation despite substantial demand. Thus he suggests that:

> It is unlikely that any amount of money devoted to inventive activity in 1800 could have produced modern, wide-spectrum antibiotics... The supply of certain classes of inventions is, at some times, completely inelastic - zero output at all levels of prices... On the other hand, the purely demand-oriented approach virtually assumes the problem [of innovation] away. The interesting economic situations surely lie in that vast intermediate region of possibilities where supply elasticities are *greater than zero but less than infinity*!
>
> (Rosenberg 1974, p. 106; emphasis added.)

This chapter explores vaccine innovation and finds that poliomyelitis is indeed a case where supply elasticities are somewhere between zero and infinity, where demand was *necessary but not sufficient*. Demand was *necessary* because unfettered research was not coming together into a vaccine without the intervention of individuals and organisations, and demand was *not sufficient* because there were failures when the knowledge required for vaccine innovation had not yet been generated.

The chapter examines how some technical problems can be shifted along Rosenberg's spectrum into more tractable and readily solvable ones. It traces the unpredictable twists and turns that characterise the evolution of knowledge, the historical circumstances in which certain research paths were taken and others abandoned, and the local social contexts into which vaccines were introduced. A recurrent theme emerges from this kind of analysis; technologists must be able to generate knowledge that is reliable, robust and shared, in order for it to accumulate and yield innovations. Moreover, there is an identifiable system underlying this process; technologists skilfully test ideas with instruments[1] under varying conditions, according to widely accepted standards, and with the active participation of coordinating institutions. This triad of elements - testing through stepping stones and intermediate conditions, skills and instruments, and policy institutions - allows us to frame historical experiences in vaccine innovation as a learning process centred on the accumulation of technological knowledge.

This chapter proceeds in seven sections. The introductory section explains why poliomyelitis drew attention and became recognised as a problem (Rosenberg 2002). It describes the construction of a vision (Blume 1992, pp. 64–70). The vision was an articulation of demand led by: a US President who championed the need for intervention; scientists who established pathology and epidemiology – how the disease is caused and spread (Thagard 1999); and technologists who conceived the idea of a potential vaccine and its operational principle – how a technology works (Vincenti 1990, p. 209). However, a vision is not sufficient for vaccine innovation; the second section describes some of the failures and barriers faced by technologists, which might otherwise be overlooked in a history of poliomyelitis successes.

The third section describes the development of a testing regime (Yaqub 2008), which involved improving learning conditions (Nelson 2008; Nightingale 2004), instrumentalities (Price 1984), and knowledge coordination and integration (Chataway *et al.* 2007). The fourth section describes the development of field-based testing - clinical capabilities - to manage the move out of the laboratory safely. The fifth and sixth sections follow the development of two trajectories with different operational principles (Dosi 1982), disagreements between people working on the different approaches, and the significance of ethics and path-dependency to certain testability issues. A final section concludes the chapter.

The emergence of a problem and the vision of a solution

Poliomyelitis was eventually given its name as a disease in the middle of the nineteenth century[2] after physicians associated paralysis with inflammation (*itis*) of the grey (*polios*) matter of the spinal cord (*myelos*) of children. Initially associated with teething, the clustering of cases in households suggested an infectious disease (Carter 1965, p. 8). This was shown in 1908 when Landsteiner and Popper infected monkeys by inoculating their brains with infected human spinal cord tissue (Robbins 2004, p. 17). The following year, Flexner and Lewis passed the infection between monkeys (Carter 1965, p. 9) starting the search for the infectious agent (Mullan 1989, p. 100).

Since infectious disease agents can be poisoned (with drugs) or killed by an immune system (primed by vaccines) this diagnosis added new potential options for public health strategies, which moved away from environmental improvement and quarantine towards prophylactic vaccine development (Baldwin 1999; Tomes 1990; Hortsmann 1985). However, the exact operational principle – how the vaccine would work - remained unclear without further investigation of the epidemiology and pathology of poliovirus.

By 1910 it was demonstrated that monkeys surviving poliomyelitis resisted re-infection and their blood contained a substance that neutralised the virus (Paul 1971, p. 108). This was a major 'proof of concept' finding. As a result, in 1911 Flexner issued a press release predicting a remedy within six months (ibid, p.116 and p. 125). However, several obstacles remained. Laboratory diagnosis was dependent on testing spinal fluid, obtained through a specialised, painful and dangerous procedure, with the result that serum was scarce and unreliable (Rogers 1992). Vaccines' operational principles depend on safely stimulating an immune response by manipulating and developing the virus to limit pathogenic (disease-causing) qualities, while accentuating immunogenic (immune-response stimulation) qualities. In general, the more pathogen there is, the larger number of substances and techniques that alter the pathogen can be explored to determine whether the changes produce a 'safe' and 'appropriate' response. It was quickly recognised that the lack of sufficient wild-type virus was a significant barrier to understanding poliomyelitis pathogenesis and to developing a vaccine (Robbins 2004, p. 17).

It is now known that three types of poliovirus can enter the mouth and nose in saliva droplets or microscopic faeces and then reproduce in the gut. Normally the immune system limits infection, but in 1–2 per cent of cases the virus travels through the blood into the central nervous system, causing meningitis and paralysis (Racaniello 2006). The model postulated by Flexner in 1913 was different (Rogers 1992) and suggested poliomyelitis travelled through the sinuses to the brain and spine, and grew in nervous tissue. These assumptions led researchers down three dead ends. First, they tried to culture the virus in nervous tissue. Second, they reasoned that vaccines would not work because poliovirus did not enter the bloodstream. Third, they were unaware they were dealing with multiple types of virus, which made comparing experimental results across and within laboratories difficult. As a result, the consensus of the scientific community from 1913 until 1935 was that a vaccine was possible but unlikely (Carter 1965, p. 58; Paul 1971, p. 113).

By 1916, the annual incidence of paralytic poliomyelitis in the US was over 27,000, killing more than 7,000. Reported cases had never exceeded 7.9 per 100,000 before but, in 1916, the rate jumped to 28.5 per 100,000 (Paul 1971, p. 148; Rogers 1992, p. 10). Hospitals refused to admit new cases, cities began insect control programmes and impounded cats and dogs (Paul 1971, p. 291) and parents sealed windows and refused to let children play outside (Oshinsky 2005).

When Franklin Roosevelt was struck by poliomyelitis in 1921, his carefully handled public relations altered perceptions of disability and helped boost scientific research (Gallagher 1985; Oshinsky 2005).[3] When it was reported in 1924 that he bathed in Warm Springs Georgia, other sufferers followed. Roosevelt spent two thirds of his personal fortune turning it into the Warm Springs Foundation under the direction of his former law partner, Basil O'Connor, whose daughter later died from poliomyelitis (ibid).

In 1934 Roosevelt staged nationwide charity balls 'to dance so others may walk' (Carter 1965, p. 14) to relieve Warm Springs' debts (Rose 2003). Despite the stock market crash the campaign raised $1m that year, $0.75m the next, and reserved $100,000 to 'stimulate and further the meritorious work being done in the field of infantile paralysis' (Carter 1965, pp. 14–18). The first 16 research grants totalled $250,000, one of which, for $65,000, was distributed to Maurice Brodie (Benison 1967, p. 179).

Brodie-Kolmer vaccine failures: a weak testing regime in need of strengthening

The optimism of the time was based on the successes of tetanus and diphtheria vaccines, which had saved millions of lives by 1910 (Chase 1982, p. 302). They were based on passive immunisation, using sera drawn from the blood of immunised horses. Flexner and Lewis attempted to replicate this approach but reported 'failure to produce neutralising serum in the horse ... [it] displayed no power whatever to inhibit the action of the virus' (ibid). Poliovirus could not be grown in horses, or in any other non-primate. Further research therefore required either humans or monkeys - 'cranky, expensive creatures, which (prior to antibiotics) had a way of succumbing to other diseases before the researcher could measure its responses to poliomyelitis. No laboratory combined sufficient interest with enough funds to maintain all the monkeys needed for thorough study of poliovirus' (Carter 1965, p. 19).

Had it been possible to infect mice or rats, they could be used as cheap, fast and simple animal models to generate experimental data on infection (Nightingale 2000). Since this was not possible innovation took a path of 'testing as validation'. In 1936, two rival investigators independently conducted field trials of vaccines (Chase 1982; Robbins 2004). Brodie and Park used a formalin-treated preparation of mashed-up infected monkey spinal cord (i.e. a killed virus approach), while Kolmer used 'a veritable witches' brew' (Paul 1971, p. 258) of live virus made from spinal cords treated with chemicals and refrigeration. The two teams hurried their vaccines into trials. Many of the 12,000 children Kolmer vaccinated were killed or paralysed, which sparked a two decades' 'wave of revulsion against human vaccination' (Paul 1971, p. 260).[4]

The trials were indicative of the community's testing norms (Constant 1980, p. 8). Vaccinology was seen as an empirical science that did not require knowledge of how a vaccine worked. Smallpox and rabies vaccines had been developed without formal identification of their infectious agents. This approach of 'try it, and see if it passes validation tests' is fine if the vaccine works. But failures provide researchers no additional insight. A fundamentally different kind of testing is required; a style where experiments actively create new conditions, a more sophisticated testing regime.

In hindsight, testing in this case was impeded in three ways. First, feedback loops in learning cycles were weak because so little virus was available. Few researchers could diagnose infection quickly by extracting spinal fluid, so most researchers had to wait for symptoms when testing for any immunity. Second, experimental iterations and refinements could not be made cheaply, easily, or quickly because of the lack of a simple model. Monkeys are difficult, slow and expensive to work with. Brodie only tested his vaccine on 20 monkeys before testing 300 children while Kolmer tested a few monkeys, himself, his children and 22 others before distributing his vaccine (Paul 1971). Thirdly, the community did not establish the types of poliomyelitis virus they were working with before trialling. Each issue needed to be addressed.

The failures moved Roosevelt to abandon the Birthdays Balls after 1937 and rename the Warm Springs Foundation the National Foundation for Infantile Paralysis (Rose 2003). Its mission was to 'ensure that every possible research agency in the country is adequately financed to carry out investigations into the cause of infantile paralysis and the methods by which it may be prevented' (Carter 1965, p. 15). Significantly, it would also '*lead, direct, and unify* the fight of every phase of this sickness' (Markel 2005, p. 1408 emphasis added). Radio promotion raised over $1.8m in a week, with proceeds increasing so that by 1945 receipts totalled $18m, and by 1955 $67m (Carter 1965, p. 26). Between 1938 and 1962, the Foundation's overall income was $630m with 11 per cent ($69m) spent on vaccine R&D (Paul 1971, p. 312).

The construction of a more sophisticated testability regime

This section describes how a new testing regime was triggered by at least three distinct policy interventions. In 1947, Harry Weaver was appointed as Director of Research to manage the research effort. He invited leading researchers to conferences, published proceedings (Smith 1990) and instituted round table discussions to 'encourage communication and intellectual cross-fertilisation in a field notable for its lack of both' (Carter 1965, p. 57). These led Weaver to the view that while undirected researchers establish more certainties about a disease, they often investigated questions of little technological relevance. He wrote to O'Connor:

> Only an appalling few ... were really trying to solve the problem of poliomyelitis.... If real progress were to be made, more exact methods of research would have to be clearly defined, procedures and techniques would have to be developed ... individual groups would have to sacrifice ... their inherent right to roam the field, and concentrate their energies on one, or at most, a few objectives.
>
> (Carter, 1965, p. 57.)

Previously the Foundation funded investigator-initiated projects (Benison 1967; Smith 1990), but Weaver set up a Scientific Research Committee to more carefully direct and coordinate research. Its head, Dr Thomas Rivers, noted:

... the Scientific Research Committee received any number of applications from individual investigators and, while many were worthwhile in themselves, together they did not seem to be going anywhere. They were too haphazard for a program and I thought that the Foundation would be better served if a committee surveyed the field and blocked out problems that needed solution ... the committee should seek out the men and institutions capable of researching such problems and support them.

(Benison 1967, p. 231.)

Despite resistance, Rivers and Weaver directed an 11-point research plan that tackled the three impediments to vaccine development directly (ibid, p. 229). After Weaver complained that experiments were 'botched by scientists who used too few monkeys or made the error of reusing monkeys whose systems were misleadingly immune to another type of the virus' O'Connor resolved to 'go into the monkey business' (Carter 1965, p. 73).

Establishing Okatie Farms

As noted previously, after decades of trying, researchers had not infected small, inexpensive laboratory animals. Since economies of scale in experimentation, unlike production, typically depend on reductions in size (Nightingale 2000) this substantially constrained research. While monkeys could be infected they required specialised care and quarters (Paul 1971, p. 101). Researchers had to spend significant proportions of their time arranging their housing and feeding, and 'placating the assistants who had to work with them' (Smith 1990, p. 123). Skilled technicians were needed to clean, feed, look after, and handle the monkeys while contending with the risk of bites, thumps and disease.[5]

Despite these difficulties, demand outstripped supply as capturing wild monkeys was difficult. Cynomolgous monkeys provided good disease models and were easier to work with, but they were scarce and expensive to import from the Philippines and Indonesia (*Time* 1954; Smith 1990). Rhesus monkeys were more abundant in India, but are sacred to Hindus. Researchers complained that their monkeys arrived dead or diseased (see Salk's correspondence in Carter 1965, p. 75).

To address these problems O'Connor established Okatie Farms while Weaver organised monkey 'airlifts' from India and Indonesia (*Time* 1954, p. 7). At the Farms monkeys recuperated before being dispatched to laboratories, saving laboratory time, effort and space. Smith (1990, p. 121) describes the Farms as 'a rehabilitation facility that was also a centre for research in the solution of problems nobody else much cared about'. The Farms developed carefully formulated dry monkey-feed (Carter 1965, p. 76) and provided instructions on feeding (Smith 1990, p. 122).[6] The historical record contains long correspondences regarding the minutiae of delivering, feeding, handling and disposing of monkeys (Carter 1965). This 'monkey business' substantially lowered the costs, improved the quality and raised the comparability of experimentation. Once the monkeys were in place the next barrier could be addressed.

Tissue culturing

The effort to develop better methods for propagating the virus continued unsuc-cessfully throughout the 1930s (Robbins 2004). By 1940 two groups grew polio-virus in human embryonic brain tissue, but did not extend the technique to non-nervous system tissues (Burnet and Jackson 1940; Sabin and Olitsky 1936). Robbins (2004) suggests this delayed the vaccine by almost a decade.

This reflected the orthodoxy that poliovirus was a nervous system virus, which occasionally spilled over into the blood. Unfortunately the finding that the virus could grow in brain tissue only reinforced the notion that poliovirus was neuro-tropic. It was therefore thought that a vaccine was dangerously impractical because it was impossible to remove all the monkey-nerve cells during prepara-tion, raising the risk of fatal encephalitis (Rogers 1992).

However, Paul and Trask discovered the virus in human faeces, implying it could reproduce in the alimentary tract (Paul 1971, p. 281). In 1940, Bodian and Howe infected chimpanzees by feeding them and in 1947 Melnick and Hortsmann demonstrated the animals' resistance to re-infection (Paul 1971, p. 287), strongly indicating an intestinal infection.

The Foundation commissioned several groups working on culturing techniques and supplied them with poliovirus and funding (Carter 1965, p. 60; Chase 1982, p. 292). The Foundation also helped source embryonic tissue from local mater-nity hospitals and proactively dealt with the qualitatively different social concerns that arose from experiments with human tissues derived from the foreskins of new-born boys, placentas, miscarriages and still-born tissue (Smith 1990).[7] A major breakthrough came when Enders, Weller and Robbins succeeded in culti-vating poliovirus in human non-nervous tissues (embryonic skin muscle). Soon poliomyelitis was propagated in a variety of tissues (Robbins 2004, p. 18).

While initially underappreciated (Paul 1971, p. 373), tissue culture trans-formed testing by providing a safer and simpler experimental environment, tighter feedback loops and faster testing cycles, so that experimental knowledge could accumulate faster. It drastically reduced the need to import monkeys, which saved money and time. As Chase (1982, p. 286) notes 'worse than the costs of buying and maintaining these animals were the temporal limits they placed on the investigative progress'. With the need for experimental animals vastly reduced, more ideas could be tested, costs reduced, feedback loops shortened, and results could be assessed more quickly (Nightingale 2000).

Tissue culturing also provided better quality virus that was relatively free of protein and, crucially, free of encephalitis-causing nerve cells (Robbins 2004, p. 18). This reopened a previously closed operational principle. The Foundation exploited the breakthrough by educating specialists and training technicians so the technique would diffuse quickly and continue to be developed (Carter 1965, p. 26). This 'tinkering' knowledge is important because to achieve good yields, cultures have to be kept at precise temperatures, in very clean containers, of the right shape and size, with the right kind of lids and stoppers (Smith 1990). The technique was improved in 1953 when human embryonic tissue was substituted with the testicles or kidneys of monkeys, generating a substantial economy of scale. A single testicle or kidney could provide enough tissue culture for two hundred test tubes when the previous method generated enough for one (Carter 1965, p. 114).

Finally, tissue cultures could be used to establish shared standards. Since infected cells were rapidly destroyed (Chase 1982, p. 292; Robbins 2004, p. 19) the cytopathic effect could indicate viral replication and the presence of viruses. With some technical modifications, tissue cultures were also used for virus titration, antibody quantification, virus isolation from clinical specimens and antigenic typing of virus isolates (Robbins 2004, p. 19).

These improvements allowed research groups to compare their results, which then revealed new paradoxes as similar experiments were producing different results. As Robbins (2004, p. 18) reflects, small differences in how long or how often nutrient media were changed were enough to produce opposite results. Rivers sought to find flaws in earlier research to explain the different results (Carter 1965, p. 90) but realised that different research groups were using different viruses. Sabin and Olitsky's experiments had failed because their MV virus was the only poliovirus that would not grow in non-nervous tissue. Rivers noted that if they 'had worked with another strain … the chances are that … we would have had a breakthrough' much earlier (Carter 1965, p. 91). Working without a clear catalogue of the various poliomyelitis strains had impeded vaccine development because the scientific facts that were established were local to particular virus types, which made comparisons difficult.

Virus typing: a 'dull and menial' programme

In 1948, Weaver pushed virus typing as an important, but theoretically unexciting, strategic research project. For a long time it was suspected that multiple strains of poliovirus existed but to establish this would involve a longer systematic effort that would involve substantial investments in laboratory space, monkeys, technical personnel and equipment. Much like the Human Genome Project, senior researchers were reluctant to take on the 'drudgery' of several years of mechanical and boring work (Carter 1965, p. 61).

Immunological testing was difficult, imprecise and time consuming. A group of monkeys was infected with a strain of poliovirus, say Type I virus, and allowed to recover. If they got sick when challenged with 'standard' doses of an unknown virus, one infers a new strain, say Type II, is present. This strain can then be injected into another group of monkeys that have recovered from Type II virus infection. If they remain healthy, the unknown strain can be confirmed as Type II. If they get sick the procedure is repeated.

The whole protocol, even when executed perfectly and with a lot of luck, would have required many monkeys. But there are many inaccuracies in making the deductions. Preparing 'standard' challenge stocks is a delicate, time consuming and frustrating job because viruses differed in their pathogenicity and infectivity. As a result, the standard dose was different for each virus strain and could be miscalculated easily. When challenge stocks are too weak, very mild infection can be mistaken for prior immunity. When they are too strong the monkeys end up dead. To guard against such miscalculations, each step of the process needed to be repeated with dozens of monkey groups (Smith 1990). Only then can a public challenge-stock database be compiled and shared.

Weaver initially set up an eminent advisory committee to lead the project but they were uninspired. Jonas Salk, who was to develop a working vaccine, had just set up a new laboratory of his own after having worked on a formalin-inactivated influenza vaccine with his mentor, Thomas Francis, for the US Armed Forces (Carter 1965, p. 35; Galambos and Sewell 1995, p. 47). Salk was looking for his laboratory's first grant when he was encouraged to take on the work being offered by the Foundation (Carter 1965). It was seen by Salk as 'a dull but dependable investment that would provide a regular dividend of money' (Smith 1990, pp. 110–117).

The large scale experiment spanned four universities and two years, acquired and classified over 200 clinical strains of poliovirus, cost $1.37m and used 30,000 monkeys imported at great expense (Chase 1982). To put this in perspective, only 17,500 monkeys had been used in all previous experiments (Carter 1965; Chase 1982), and in 2002 only 52,000 non-human primates were used across the entire US R&D system. The project showed conclusively that there were three, and only three, immunologically distinct types of poliomyelitis virus (Bodian 1949; *Time* 1953).

This crucial information provided a standard against which future vaccine candidates could be compared (Carter 1965, p. 275). With the chances of making a poliomyelitis vaccine much improved, a number of groups worked towards that goal but with different operational principles underlying different trajectories. Hammon chose to pursue a passive immunisation approach, while Salk and Sabin successfully pursued active immunisation approaches[8]. Salk took the line of a formalin-inactivated vaccine, while Sabin chose to pursue a live attenuated vaccine.

Passive immunisation: testing for design and field-based capabilities

By the 1950s, the emphasis shifted from establishing proof of concept in monkeys and learning about how the vaccine would work in monkeys (the operational principle), to facilitating learning by actively creating and controlling conditions in humans. These learning conditions would be more relevant and realistic than the conditions created in monkeys, but they would bring associated complexity and be harder to create and control in humans. This section highlights how the complexity was managed so that testing resources were coordinated and tests on humans resulted in the accumulation of technological knowledge.

A critical part of the vaccine design process can be described as a difficult and uncertain transformation of qualitative goals into objective ones. I begin by outlining the feasibility of passive immunisation as an operational principle, before analysing considerations made about vaccine design and organisational capabilities during the move to human testing.

Hammon believed that gamma globulin, an antibody obtained from pooled plasma with known neutralising activity in the laboratory, might offer benefits in practice. Rather than prevent poliovirus infection, his immediate goal was to prevent infection-causing disease on the nervous system (Carter 1965; Paul 1971; Plotkin and Vidor 2004). Permanent immunity through repeated infection might be achieved, but without the symptoms of poliomyelitis. The idea carried weight in part because passive administration of serum achieved some success against measles virus (MRC 1948).

In 1948 Morgan and Bodian were able to protect monkeys from one type of poliomyelitis (Carter 1965, p. 64; Paul 1971, p. 405). By using graded doses of virus with the purpose of producing varying levels of antibody, they effectively constructed an index of the degree of immunity in monkeys. This represented an improvement in the knowledge infrastructure because future antibody experiments conducted by different research groups could be compared to a shared index. Hammon argued that the role of antibody was still uncertain in humans, and that the antibody index would allow a safe start to ascertaining 'how much was enough for humans?' and 'how long do they last in the blood?'.

The Foundation created a 'Committee on Immunisation' to manage strategic and logistic aspects of human vaccine trials (Carter 1965, p. 125; Paul 1971, p. 407). It was a daring role given the traumatic failures of the Brodie-Kolmer trials two decades earlier. Fear about using killed or live virus was a common theme voiced by Sabin and coloured the views of most in the field (Rinaldo 2005). However, Hammon's vaccine did not contain any virus and answered Rivers' call for boldness, 'I think it is time that we got ready to go somewhere, and somebody ought to come up with some concrete experiments that will be done in human beings on a small scale in order to get going' (Carter 1965, p. 126).

Hammon's preliminary field trial showed that relatively low levels of antibody could prevent invasion of the central nervous system (Hammon et al. 1953). The results provided vaccine developers pursuing different trajectories, such as Salk and Sabin, not only with the confidence that disease could be prevented, but also a tangible performance criterion. The subjective aim of immunity had become an objective goal of putting antibodies in the blood.[9] The testing regime had a bar against which potential designs could be compared.

Questions remained of how quickly and safely immunity could be established in the blood, and how long it would last for in the blood under various conditions. Antibodies produced by the body through active stimulation were thought to last longer than those passively given to the body. Hammon argued the other non-passive trajectories had safety concerns and needed multiple injections, saying that with his gamma globulin, 'its effect would be immediate and would represent no danger to any child' (Hammon 1950, p. 702). Although passive immunisation might not need multiple injections, Hammon apparently overlooked the fact that his subsequent clinical trial would seriously deplete all reserves of gamma globulin.[10] A further trial with more people, and hence more slightly varied conditions, was needed to address these issues of speed, durability and quality of immunity.[11]

The Foundation funded Hortsmann (1952) and Bodian (1952) to see if passive immunisation protected monkeys from very high, lethal doses of poliovirus of all three strains. Compared with Morgan's experiment in 1948, these conditions were more technologically relevant (because they involved all three strains), and perhaps even scientifically less interesting (because the theoretical concept of neutralising antibodies had already been established). The protection achieved under these conditions convinced the panel to fund a pilot study of 5,000 children. Panel members realised that this size would not yield statistically significant results, rather the study's purpose was 'to gain experience in organisation and administration, as well as to evaluate the public's and medical profession's reaction to such a trial' (Rinaldo 2005, p. 793).

The details of the trial that needed to be organised were very broad.[12] Most critical was 'the definition of the severity of the paralytic disease, for which they used a carefully graded scale of muscle function loss' (Rinaldo 2005, p. 793). Similar to the virus typing project, and the antibody index, this is another example where the Foundation set up an infrastructure to compare future observations to a set of known conditions, thereby ensuring that those observations would contribute to the cumulative growth of technological knowledge. It might otherwise have been seen as a chore, with little, if any, scientific merit.

The pilot results were encouraging and public support was very strong, with hundreds of volunteers being turned away by day four (Hammon *et al.* 1953). Problems included such issues as lack of access to large autoclaves to sterilise the syringes and needles. A larger trial was quickly approved, which involved 55,000 children. The result of this trial was considered, 'conclusive evidence of a very significant reduction in the total number of cases of paralytic poliomyelitis' (Hammon *et al.* 1953, p. 758).

Hammon concluded that, 'perhaps the greatest contribution of the gamma globulin trials ... demonstrated that a very low concentration of antibodies will protect man' (Hammon *et al.* 1953, p. 1283). Aside from taking this design standard from monkeys and establishing it in human conditions, a graded scale of paralytic disease was also developed. The trials were seized as an opportunity for the Foundation to build up organisational capabilities in acquiring local knowledge for testing outside laboratory conditions, and coordinating people, resources, logistics and public support. Alongside the accumulation of technological knowledge, the Foundation had begun setting up the organisational decision-making process for moving active – but potentially more dangerous - vaccines to trial in humans.

Although much was gained through the development of the passive trajectory, it was these other potentially more dangerous active vaccines that started to look more attractive as they were developed because no matter how well a passive vaccine prevented disease progression, it would never be able to prevent infection in the first place (unlike an active vaccine).

Killed vaccines: testing regimes for taking 'calculated risk'

This section discusses how a more risky vaccine was tested in humans. The vaccine was more risky than Hammon's because it contained killed virus, but less risky than using live vaccine. The initial risk appears to have been borne by certain sections of society, who provided the conditions that were both relevant for technological development and suitable for learning. The resulting knowledge growth was shared and cumulative, as institutions put in place mechanisms to mediate differences in opinion, and coordinated carefully designed testing processes to manage the leaps from laboratory to animal and human settings.

By 1953, Salk had shown that poliovirus could be inactivated by formaldehyde (Salk 1953). Moreover, he determined how much formalin affected inactivation and conducted safety and immunogenicity studies in animals (Benison 1967; Robbins 2004). If there was any doubt as to whether such animal findings could be transferred to children, Howe's (1952) paper, entitled 'Antibody response of chimpanzees and human beings to formalin inactivated trivalent poliomyelitis vaccine', made it clear.[13] Salk, too, had started preliminary studies in humans that showed that antibodies could be increased to relatively high titres in children already infected at the Watson Home for Crippled Children.[14] But these advances, aside from any modern-day ethical testing concerns, were leading to a somewhat problematic vaccine.

Conventional wisdom held that only a live-attenuated vaccine could confer long-lasting immunity because it more closely mimicked a true infection (Carter 1965; Klein 1976; Smith 1990). Several of the Committee's senior virologists, such as Albert Sabin and the Nobel Laureate John Enders, questioned the relevance of antibodies and doubted the safety of a vaccine prepared from virulent poliovirus, regardless of how thoroughly it was inactivated, especially after the failed vaccines of the 1930s (ibid).[15] Enders is even quoted as having confronted Salk and calling his work 'quackery' (Carter 1965, p. 88). But for Salk the notion that only natural infection or a vaccine made of living pathogen could offer durable protection was nonsense (Carter 1965; Klein 1976; Smith 1990).[16]

Members of the Foundation acknowledged 'sharp differences' in the Immunisation Committee and tried to manage them (Paul 1971, p. 407). For example, regarding concerns about whether an inactivated poliomyelitis vaccine really was inactivated, Rivers said at a Committee meeting, 'I think we will all admit that there is no *test* to be sure the stuff is inactive. Why not just accept that? Why kid ourselves? Why use the word inactive? Why not just say, "safe for *use*?" It won't produce disease, and that's all there is to it' (Carter 1965, p. 126, my italics).

Such 'nervous brawling' stalled any kind of progress (Carter 1965, p. 129). So, in 1953, the Foundation set up a new and smaller committee because, as Weaver is quoted as saying, 'The immunization committee was not able to function with the necessary dispatch. It could get entangled for months in technical debates. Furthermore, its members were virologists and the decisions on which we needed help were not exclusively virological. The Vaccine Advisory Committee with experienced public health men … was a far more efficient group' (Carter 1965, p. 176; Paul 1971, p. 411). The need for a second committee suggests that the design of tests is not an entirely objective and technical matter, and includes broader considerations. It was also established in part to limit conflicts of interest that may arise from having competing designers playing the role of 'architect, carpenter and building inspector' all at once (Weaver quoted in Carter 1965, p. 179).

Salk recalls the arbitrary nature of deciding when and precisely what to test. 'All we had were several dozen experimental preparations, some with adjuvant, some without, some containing one type of virus, some another or a third or all three, some made with monkey tissue, some with testes, some inactivated for ten days, some for thirteen, some for twenty one' (Carter 1965, p. 130). Salk insisted, 'I don't know that we even have a vaccine yet. That term was used, but I think it should be understood that we are using it as a colloquial expression. We have preparations which have induced antibody formation in human subjects' (Carter 1965, p. 152). The possible permutations of experimental conditions Salk describes seem endless. Moving to major field trials involved uncertain but skilled judgement about what effects were likely to be reliable, and Salk was pushed into readiness by the Foundation.

Salk's uncertain vaccine was moved into trials with a sense of purpose and social conscience.[17] Rivers asked, 'Wouldn't it be silly to wait 50 years or to wait 10 years to develop the ideal vaccine when there is the possibility of a vaccine being developed very rapidly that will last, say, for two or three years with one injection perhaps? We don't know anything about that, but have we the right to wait until the ideal vaccine comes along?' (Carter 1965, p. 151). Although Rivers thought the Salk vaccine was 'something slightly better than gamma globulin, something by definition imperfectible', he felt it was 'worth a try' (Carter 1965, p. 152). From these exchanges, it seems that vaccine development is not a process of optimisation. Vaccines are developed to function only sufficiently well enough to fulfil a social purpose. That purpose drove the Foundation to begin planning for a major field trial.

Harry Weaver wrote, 'The practice of medicine is based on a calculated risk … the physician elects to follow the course that provides the greatest benefit with the least risk of incurring any untoward effects… If [we wait until more] research is carried out, large numbers of human beings will develop poliomyelitis who might have been prevented from doing so … our work must be governed by scientific and sociological concerns' (Benison 1967; Carter 1965, p. 147).[18] The vaccine development process was not simply a scientific puzzle, with a technical solution that could be found and optimised; rather, the urgency of the historical and social context of the actors played important roles in their decisions about an 'imperfectable' vaccine.

In the design of the trial, the planned use of placebo controls was problematic, but the precedent seemed necessary. Initially, Weaver sought simplicity and economy, and suggested that the poliomyelitis rate be compared between vaccinated and non-vaccinated school-children of the same age (Carter 1965, p. 176). However, the Vaccine Advisory Committee suggested that socioeconomic differences between those who volunteered and those who did not would weaken the study.[19]

Salk felt that his vaccine was not up to such a stringent test, and lapses in the manufacturing process or unimpressive results of a double-blind test might scupper the opportunity to improve it (Carter 1965, p. 178). I quote him at length in the paragraphs below to show that the design of the tests was at the centre of his concerns at the time, and that the parameters of the tests left an indelible mark on the nature and characteristics of the vaccine.

> The sensible thing, I thought, was to accept the urgencies of the situation and continue improving the vaccine. I thought the field trial should be designed to permit this, not prevent it... I thought we should concentrate on polio prevention and be less concerned about making epidemiological history with an elegant double-blind study. I was afraid that, for some people, the *kind* of test had become more important than the kind of protection the vaccine might be able to provide.
>
> I wanted to know who had been vaccinated so that blood samples could be taken promptly. If tests then showed that a certain batch of vaccine was producing unsatisfactory results, the children could be revaccinated with better material. At the same time, we could be taking steps to improve the manufacturing process and avoid new batches of inferior vaccine. Finally I was uncomfortable about giving placebo shots to children, depriving them of immunity in what might turn out to be an epidemic year. Many public health officials agreed with me on this.
>
> You had this rigid insistence that a 'product' be submitted forthwith for ceremonious testing. The emphasis on 'product' and on ritual and on looking good in the eyes of certain elements in the scientific community was being allowed to obscure the real purpose of everyone's work, which was the prevention of polio.
>
> (Carter 1965, p. 178.)[20]

In order to address the concerns of parents, teachers and such 'humanitarians' O'Connor announced that an observed control plan would be used, in which children would not be injected but only observed (Meldrum 1998). The Foundation asked health officers for advice and support, and they suggested that the Foundation may not be able to maintain impartiality in such evaluation (Meldrum 1998). So O'Connor appointed Thomas Francis to head the evaluation of the trials, a critical but unglamorous task, based on 'his deft direction of complex field trials of influenza virus vaccines during World War II' (Markel 2005, p. 1408). However, Francis would not accept until he manoeuvred between health officers, paediatricians, clinical poliomyelitis specialists, statisticians and virologists to engineer a change in the trial design (Meldrum 1998). He suggested a placebo design run in some areas at the same time as an observed design run in other areas.

Addressing concerns about volunteer recruitment in the placebo plan, the evaluation group decided that it could rely on the widespread fear of the disease; members agreed that 'it would not be difficult to sell as there is a high attack rate ... [and] there would still be a 50% chance of a child receiving the vaccine' (Meldrum 1998, p. 1235). Francis compromised with Salk and others to a certain extent with observed design in some areas, but his insistence on the placebo plans in other areas was particularly important in the context of the vociferous criticisms from Enders, Sabin and others about the validity of the killed-vaccine concept.

Firstly, results emerging from double-blind trials might be more convincing, and facilitate quicker and more widespread vaccine adoption. Secondly, it was important given the possible conflict of interest arising from the Foundation evaluating a vaccine that they, as an organisation, developed and sponsored. Thirdly, the placebo plans were also a part of the Foundation's effort to legitimise an institution governed by non-experts.

The trial for the vaccine went ahead in 1954 and was the largest of its kind to be run. It was not a cheap gamble as grants for the field trial and its evaluation cost the Foundation a total of $7.5m. The results of nearly 2 million children were presented on 12 April 1955, and the vaccine was found to be safe and 70 per cent effective (Smith 1990). Although not completely effective, the breakthrough cases[21] were judged to be less severe (Smith 1990). With financial guarantees from the Foundation, industrial production facilities were already built and ready to operate (Blume and Geesink 2000). The Foundation paid a further $7.5m to the manufacturers for 10 million Salk vaccine doses (Chase 1982). The products of six producers were licensed within days, one of whom was Cutter Laboratories in Berkeley (Offit 2005).[22] Poliomyelitis cases dropped from 58,000 in 1952 to 5,600 in 1957.

Paul, whose career in poliovirus research spanned both eras, contrasts the 1935 and 1955 vaccines. 'The situations were in no way comparable, for the Brodie-Kolmer vaccines had been launched in the face of colossal ignorance, whereas the Salk-type vaccine had been promoted under circumstances which from the start almost guaranteed success. And yet one cannot help feeling a twinge of sympathy for the two figures of 1935 who were so alone in the midst of their disgrace, in contrast to the powerful forces of the National Foundation, the US Public Health Service, and innumerable advisory committees that stood back of the Salk type vaccine' (Paul 1971, p. 420). Paul notes how different the testing regimes were and how the difference critically changed the environment from which a vaccine could emerge.

Live vaccines: testing in the shadow of the killed vaccine

This section reviews how improvements to the testing regime enabled the establishment of a live vaccine trajectory. It emphasises the historical dependency of such trajectories by highlighting the role of non-fiscal testing resources, shows how dependent technological trajectories become on context, systems and testability constraints, and ultimately describes different decisions taken by public health authorities in the USSR and USA.

As he had done with the yellow fever virus, Max Theiler passaged the poliovirus continuously through the brains of living mice until, without losing its capacities to stimulate an immune response, the attenuated virus no longer caused paralysis (Chase 1982). He reported it to the Foundation in 1946, which then funded further research to see if poliovirus could also lose its ability to infect the central nervous system – which it did on repeated passage through non-nervous system tissues (Robbins 2004).[23]

The live attenuated poliomyelitis vaccine approach was feasible only after certain developments in the testing regime because the approach relied on striking a balance between efficacy and safety. This entailed searching for virus that is not pathogenic (disease-causing) but retains some of its virulence (ability to infect). By strengthening the testing regime, the Foundation enabled safe learning in humans through variation-selection (Campbell 1960; Vincenti 1990; Yaqub and Nightingale 2012). The development of tissue culture techniques[24] facilitated the rapid emergence of variation in strains, while the availability of monkey models allowed vaccine developers to select for pathogenicity and virulence traits[25], and the typing project allowed putative vaccine preparations to be challenged without added confusion.

Sabin was one of several groups[26] working in this way (Paul 1971; Robbins 2004). In light of the improvements to the testing regime noted above, the Foundation provided him with $1.2m between 1953 and 1961, and $2m in total (Carter 1965, p. 357; Chase 1982, p. 303). In 1955, Sabin began a trial on inmates in Chillicothe Federal Prison in Ohio (Carter 1965, p. 357; Smith 1990, p. 301). His vaccine was successful, but the Foundation saw little reason to take chances with a larger scale trial of an infectious live vaccine when Salk's field trial had demonstrated efficacy the previous year. Large scale trials of Sabin's vaccine, and those of others, would be difficult to interpret because the Salk vaccine had been licensed and was being used widely. For example, when, in 1959, Herald Cox had the opportunity to test his live vaccine in Miami, Florida, Sabin dismissed any excitement by pointing out that too many people had taken Salk vaccine for the test to mean anything (Carter 1965, p. 365).

There is clearly a strong path-dependency element to testing processes in vaccines (see also Blume 2005), but I would like to draw attention to a slightly different view. In the early experiments, poliomyelitis researchers faced a shortage of virus; Evans and Green, who were beaten to the Nobel Prize, faced shortage of human embryonic tissue; Hammon faced issues with a shortage of gamma globulin; while Sabin faced a shortage of people to test on. These cases represent a scarcity of testing resources. These resources are not fiscal, as is commonly emphasised in health and vaccine development literature (see, for example, Archibugi and Bizzarri 2004; Arnold 2005; Barder 2005; Lanjouw 2003), but can be anything from the availability of monkeys, gamma globulin, primary isolates, to simply people as test subjects. They were unlikely to have been resolved by market failure approaches or policies that focussed on pecuniary issues alone.

The safety concerns in this trajectory extended beyond simply whether the virus in the vaccine was sufficiently attenuated to prevent it from causing disease. The major concern centred on its genetic stability and whether the attenuated virus would *remain* safely attenuated. One of the advantages of the live vaccine was that after it passed through the intestines and was excreted by the vaccinee, it might then go on to confer immunity to someone else in the community. But the same advantage became a disadvantage for those who thought that, after several passages through the community, the altered vaccine strain might undergo progressive genetic changes such that it reaches a degree of virulence comparable to that of wild epidemic polioviruses. The success of the entire live approach therefore turned on proving that any cases of poliomyelitis was not caused by the vaccine reverting back to virulence after replication in the host.

Melnick found that live vaccine virus passaged through children was sometimes virulent enough to paralyze monkeys (Carter 1965, p. 381). This caused serious concern, but there was no way in which a test could show that a given case of poliomyelitis in humans had been caused by the live vaccine, even if the victim was struck by poliomyelitis shortly after taking a live vaccine. If virus recovered from the victim resembled the wild type, one could suppose that it had taken over the intestines, and driven away the vaccine virus, before causing the disease (wild type-induced disease). Alternatively, one could decide that the vaccine virus had changed to resemble the wild type and become virulent, thereby causing vaccine-induced disease. Either way, testing primary isolates would not be able to prove a vaccine guilty.

This made designing a test for measuring live vaccine safety virtually impossible, never mind one that could be compared with the safety of a killed vaccine. In the absence of a test that could offer commensurable assurances of safety, the live vaccine continued to be perceived as being more risky.[27] Closely tied with these perceptions were the assumptions of the vaccine designers about the social context in which their designs would be used. Safety would become more readily observable as a systemic feature, as protagonists argued risks and benefits in different contexts.

Due to safety concerns and the prior use of Salk's vaccine, Sabin was forced to look abroad to conduct large scale trials. In 1958, 200,000 children in a Singapore trial received Sabin's live vaccine in an effort to curtail their epidemic (Paul 1971, p. 454). By 1960, approximately 100 million people in the former USSR and Eastern European countries had received the vaccine. By the end of the year enough evidence had been established to secure licensure in the US for Sabin's live vaccine (Paul 1971, p. 456).

However, the continued existence of distinct trajectories depended on variation in health systems because a given vaccine-attribute could serve as a merit in one and as a drawback in another. As a Soviet public health official remarked, 'Our inoculation program was a public-health measure, not a field trial. It was designed to suit our medical services. In attempting to inoculate a population the size of ours, could there be any serious confusion about whether to give away candy drops, when the alternative was injection requiring so much more apparatus and personnel? Our work with the Sabin vaccine must be viewed in terms of public health and not as a strictly controlled scientific experiment' (Carter 1965, p. 359).

If the Sabin vaccine could actually be shown to cause paralytic poliomyelitis, the finding would have been more significant for the US than for the Soviet Union. The Soviet Union was suffering poliomyelitis incidence rates of 94 per million (Carter 1965, p. 363), much higher than that of the US, so any vaccine that could reduce that figure faster (because it could confer immunity to the non-vaccinated too) would be allowed the deficiency of a few vaccine-caused cases. It only represented one dimension in a broader set of criteria for the health system as a whole.

The protagonists of each vaccine promoted their interests and preferred choice, but the way in which a vaccine's attributes complemented existing infrastructure and health systems is likely to have had a greater influence in determining their adoption.[28] The ensuing history of the changing relative merits and drawbacks of the Salk and Sabin vaccines has been astutely discussed elsewhere (Blume 2005; Lindner and Blume 2006; Blume and Geesink 2000). It is worth noting, however, that as incidence of poliomyelitis decreased in the US over the next 30 years, the perception of risks and benefits changed and so the choice of vaccine changed too.[29]

Discussion and conclusions

Persistent learning and innovation can have dramatic cumulative effects. About 4,000 years ago, poliomyelitis was first recognised. About 200 years ago, its clinical features were recognised as a basis for diagnosis. About 100 years ago, its infectious properties were established. About 50 years ago, not one but two vaccines were developed. About 25 years ago, eradication commitments and plans were made. A few years ago, poliomyelitis cases were confined to as little as two countries. Developing an understanding of knowledge accumulation processes can therefore provide a basis with which to retain some of our optimism and hopes for the future.

Taken together, history and theory provide lessons about what features of knowledge accumulation are specific to the time, place and technology of poliomyelitis vaccines and what lessons are more general. This chapter mapped out the process of vaccine innovation from problematisation and diagnosis, to the development of uncertain operational principles (how a technology works), and then a process of testing across a series of stages from the protected, simplified conditions of the laboratory, to the social complexity of a WHO field worker conducting a mass vaccination programme in an isolated part of the world. In the case of poliomyelitis we see that the simple extrapolation of an older operational principle based on limited understanding produced disastrous results and, following a period of confusion, shifted the research agenda towards a new, qualitatively different, approach based on better understanding of the disease and its pathology.

While in many cases where a practical technology is being sought there are good reasons to avoid testing a new version of it in actual practice until one is confident about safety, in this case this issue was paramount, and very much shaped what was done and not done. The complexity and danger of the disease meant that experiments had to be conducted on simplified models but the absence of a simple animal model of the disease meant that experimentation was extremely costly, time consuming and difficult.

The empirical evidence stressed the role of stepping stones or intermediate conditions that needed to be created and emphasised the instruments, skills, capabilities and active governance needed to move along Rosenberg's continuum. These included the provision of monkeys to reduce the costs of experiments, which were much higher because a simpler animal model was not available. The development of tissue culture techniques to provide more consistent experimental inputs, with faster, tighter and better defined experimental cycles. Also, the establishment of a virus-typing programme to clarify a dull but important question, which was extremely expensive to answer but offered little in the way of scientific novelty for ambitious researchers. These programmes were needed to coordinate knowledge production and govern the much more difficult and complex jumps between intermediate stages when a simple experimental model that would allow fast, accurate experiments was not available.

As Rosenberg (1974) alluded to, the more general lesson for research policy might be that the provision of complex technologies through the translation of basic research is not therefore just an issue of more money, or more demand, or a social culture of public acceptance. At the same time, policy-makers should resist the temptation to simply believe that any extra knowledge will be a contribution to innovation. Work has to be done to ensure experimental knowledge can be generalised (vertically) across research groups working on the same problem, and (horizontally) across different stages of the innovation process as it explores uncertain avenues and undergoes qualitative changes between the laboratory and the clinic. Thus this chapter shows why a question with which policymakers have been long grappling is so difficult to answer *a priori*: for which diseases will a forceful R&D programme most readily yield a vaccine?

The differences in therapeutic and prophylactic outputs between neglected and non-neglected disease do not therefore just show a lack of care or social concern, nor that distributed social actors lacked power or were not aligned. They also reflect a difference in Rosenbergian supply elasticity such that some products may be beyond the reach of even the largest pharmaceutical firms, until certain aspects of the knowledge accumulation process are addressed in order to make innovation more tractable. Calls for more funding are likely to find support from the analysis of this paper, but that support is not unconditional.

As previously highlighted by Sarewitz (1996), the notion that simply funding science without any additional oversight or governance will generate solutions to such problems is a myth, with both ideological and metaphysical foundations. Similarly, suggestions that the problem simply reflects inadequate incentives, and might be solved by measures like competitions and prizes, are naïve. There are clear trade-offs at work here as, under conditions of uncertainty, decentralised exploration of multiple avenues clearly has advantages in terms of search, but disadvantages in terms of governance and coordination. That our theoretical understanding of governance focuses so much on the two extremes of pure decentralised markets and pure top down centralised control suggests a major research effort will be needed to provide policy advice on the governance of research that is practically valuable to the development of socially beneficial technologies.

Acknowledgement

This chapter contains excerpts reprinted from Yaqub, O. and Nightingale, P. (2012) 'Vaccine innovation, translational research and the management of knowledge accumulation'. *Social Science & Medicine*, Vol 75, Issue 12, pp. 2143–2150. Copyright 2012, with permission from Elsevier.

Notes

1 This effect is appreciated by engineers. 'About half of the Institution of Electrical and Electronic Engineers annual list of the two hundred top innovations is devoted to testing equipment' (Constant 1980, p. 276).
2 The earliest evidence of poliomyelitis is an Egyptian stone carving depicting a man with a deformed limb dated to 1500BC (see Paul 1971, p. 15 for an image of the stone). The disease was given a series of changing names as the characterisation of the condition became increasingly specific. For example, names range from morning paralysis in 1843 to tephromyelitis anterior acuta parenchymatose in 1872 and, later, infantile paralysis (Paul 1971, p. 5).
3 At the time afflicted individuals were often hidden and an influential orthopaedic text supported the idea that disabilities are punishments from God, stating, '...a cripple is detestable in character, a menace and burden to society, who is only apt to graduate into the mendicant and criminal classes...' (Longmore 1987, p. 357).
4 In a public health association meeting, Kolmer is reported to have said 'Gentlemen, this is one time I wish the floor would open up and swallow me' (Chase 1982, p. 284; Paul 1971, p. 260). Brodie may have killed himself four years later (Paul 1971, p. 261), but his death certificate cites thrombosis (Chase 1982, p. 284).
5 The possibility of contracting disease from monkeys was frightening enough for Salk to request from the Foundation 'a $10,000 life insurance policy for each of the individuals in this extra-hazardous work' (Carter 1965, p. 75). One physician working with monkeys suffered a fatal case of encephalitis.
6 Among the tips were to divide the rations so that the monkeys would not have enough in one go 'to fling around and mash into each other's ears and stuff down drains and such like' (Smith 1990, p. 122).
7 Embryonic tissue grows much faster than that of adults, and is less prone to disease, so it is the preferred tissue for culturing with. Given the contentious nature of acquiring such tissue, the Foundation often could only acquire small quantities of foreskins from circumcisions of new-born boys. Placentas, miscarriages and still-born tissue were also used but their supply was even less predictable (Smith 1990). Knowledge generated from dead babies has qualitatively different social contexts than knowledge generated from dead animals.
8 Passive immunisation refers to injection of blood gamma globulins that transfer specific antibodies to the virus, in contrast to active immunisation, in which an antigenic substance is injected that induces specific antibodies to the virus.
9 By helping to ascertain how much antibody was needed to prevent infection, Hammon effectively provided what can be called a correlate of immunity. Many HIV vaccine researchers lament the lack of correlates of immunity (see, for example, Garber *et al.* 2004, p. 398).
10 The limited availability of gamma globulin restricted its use. Obtaining gamma globulin was an expensive and time-consuming process and depended on voluntary blood donations. At the same time, the Korean War and hospital needs were drawing on supplies. O'Connor warned that there was not enough to provide 'even temporary protection to the 46 million children and adolescents most susceptible to poliomyelitis'

(Rinaldo 2005, p. 795). Nevertheless, the Foundation spent $7m boosting gamma globulin production and a further million children were protected in the poliovirus season of 1953 (Rinaldo 2005).

11 The Immunization Committee initially turned down Hammon's request for a larger scale controlled trial (Rinaldo 2005). They wanted to see more animal and human data before embarking on a complicated and expensive clinical trial (the trial ultimately cost the Foundation $1m). They were also concerned about using placebo controls, which had never been used before, and its moral and social acceptability (Rinaldo 2005).

12 They included how to: blind the vaccine vials, select a type of control inoculum, source and set dosage of gamma globulin, types of syringes, packaging, venue, injection administration site on the body, consider legal aspects such as written informed consent, select geographical areas undergoing epidemics of a suitable magnitude, gain approval by local population, manage publicity and preparation of clinics, and follow up studies to identify incidence cases.

13 Howe tested six children at the Rosewood school, whom he noted as 'low-grade idiots or imbeciles' (Howe 1952, p. 265), and was able to report that 'both children and chimpanzees develop readily demonstrable neutralising antibodies at comparable levels following the injection of small quantities of clarified monkey cord suspensions containing formalin inactivated poliomyelitis virus' (Howe 1952, p. 265).

14 Like Howe, Salk tested children who were not infected but were 'mentally retarded' and found that the levels of antibody production were equally encouraging (Chase 1982).

15 Enders cautioned, 'the ideal immunising agent against any virus infection should consist of a living agent exhibiting a degree of virulence so low that it may be inoculated without risk' (Enders 1954, p. 88). Sabin persistently objected that a massive investment of time, money and public faith in a [killed] vaccine of only temporary use would hurt efforts to find a live virus that would really solve the problem (Smith 1990). Flexner declared that 'only an infectious vaccine compounded of living virus could protect' (Carter 1965, p. 86).

16 It is likely that Salk's deviation from the orthodoxy resulted from his newness to the field of poliomyelitis prophylaxis. In fact, his previous experience in developing inactivated influenza vaccine most probably directed his choice of approach to the poliomyelitis problem (Galambos and Sewell 1995, p. 47). Brodie and Kolmer had tarnished the killed approach and Salk was careful in his relations with the public to set apart his methods from theirs; for example, a *Time* magazine article pointed out, in unusual technical detail, that Salk's vaccine used purified mineral oils to hold the vaccine in the body for longer as a way of distinguishing his from previous efforts (*Time* 1953).

17 A member of the Vaccine Advisory Committee said, 'I think … progress can be made even in the light of the fact we have so little knowledge. It would seem to me the time has come to really go at the inactivated material … The live virus is fine, but if you think about it as a public health measure, it is a difficult thing to use… I don't think you have a good excuse morally to go into infectious material until we have shown that inactivated material was unsatisfactory' (Carter 1965, p. 128).

18 The trade-off was emphasised in a telegram to convince sceptics, 'It is said that to await certainty is to await eternity' (Smith 1990, p. 295).

19 High income, well-educated families were more likely to submit their children to experimentation of this kind. In contrast, less well-educated families living in poorer areas were less susceptible to paralytic poliomyelitis, tending to contract the non-paralytic form in infancy and gaining immunity. Such a project would only immunise the children most susceptible. The poliomyelitis rate in the vaccinated group might be similar to that among the unvaccinated. Therefore the vaccine might be good, but the test would not have the resolving power to prove its efficacy. In addition, poliomyelitis diagnosis was still difficult despite the scale developed in the Hammon trials, and any

biases emerging from knowing who had been vaccinated and who had not would serve to exacerbate the problem.

20 Salk went on to describe the use of placebos as 'a fetish of orthodoxy ... a beautiful epidemiologic experiment over which the epidemiologist could become quite ecstatic but would make the humanitarian shudder and would make Hippocrates turn over in his grave ... the worship of science involves the sacrifice of humanitarian principles on the altar of rigid methodology' (Carter 1965, p. 192).

21 Cases where volunteers are diagnosed with poliomyelitis despite being vaccinated in the trial.

22 The Cutter incident represented 'one of the worst pharmaceutical disasters in history' (Offit 2005, p. 1411). In a batch of Salk vaccine manufactured by Cutter, there remained some virus that had not been killed. It caused over two hundred cases of poliomyelitis, of which 150 were paralytic and 11 were lethal (Nathanson and Langmuir 1963). The error paralysed 15 times more children than the earlier Brodie and Kolmer vaccines combined.

23 The idea of attenuating the poliovirus, rather than killing it outright, appealed to many because it was presumed to mimic the natural situation more effectively, resulting in a longer and more effective immunity. However, Salk, and other proponents of killed vaccine, resisted the notion that immunity provided by live vaccine would be somehow longer lasting. 'One cannot say how long immunity may last, one can report only how long it has lasted' (Carter 1965, p. 377).

24 Viral culture techniques were significantly improved by Dulbecco and Vogt (1954). Adapting techniques for growing bacteria, they grew virus in microscopically thin mono-layers of chick embryo tissue cells. The colonies proliferating from the growth of a single viral particle could be identified, counted and isolated. This made it easier to purify specific lines of virus, which was extremely valuable for those looking to prepare a live vaccine (Paul 1971, p. 406; Robbins 2004, p. 19).

25 Selecting strains with monkeys meant that live vaccine development did not need to rely on few and imprecise in-vitro markers of virulence, such as growth at higher temperature (Paul 1971, p. 458). Instead, a more authoritative test for neurovirulence, adopted as the standard by the regulatory agencies, was devised where monkeys had to be inoculated through their central nervous system (Robbins 2004, p. 20).

26 Other groups were led by Hilary Koprowski at the Wistar Institute, Herald Cox at Lederle Laboratories and Joseph Melnick at Yale, all of whom tested their prototype live vaccines on institutionalised children (Chase 1982).

27 Tommy Francis played down the difficulties of designing comparable safety tests. 'The two outlooks are, then, simply, this. Inactive virus is apparently a test of the straightforward hypothesis that antibody induced by the administration of antigen can provide protection without subjecting the recipient to harmful effects of even apparent infection. The other, through the use of modified active virus, seeks to induce antibody formation but wishes to add some undesignated advantage derived from assumedly harmless infection (I am not certain that any significant infection may not create undesirable tissue reactions...). Which of these approaches to poliomyelitis will be the more effective is, then, not a decision to be arrived at by authority and debate but by ... making the observations. When the conditions are appropriate, tests should be made.' (Carter 1965, p. 357.)

28 Cox was benefiting from an aggressive publicity campaign by Lederle touting the advantage of a single dose vaccine that still protected against all three strains (trivalent) (Carter 1965, p. 365). Koprowski managed to trial his vaccine in 9 million people but had his vaccine turned down by the US government because it caused some lesions in monkeys (Paul 1971, p. 454). Salk argued that his vaccine was effective and that they needed to wait longer, without introducing other vaccines, to see definitive results of an imperfect vaccination programme. Sabin's appeared to be the newer more modern

vaccine with which the public health service could have a second chance of executing a vaccination programme of more complete coverage (Carter 1965, p. 372). Sabin's field trials in the Soviet Union were so effective that they were doubted, and it took a report by Hortsmann, who was dispatched there by the WHO, to verify the standards and evidence (Paul 1971, p. 455; Robbins 2004, p. 20).

29 'The conclusion [of a comparative analysis of live and killed vaccine] is heavily dependent on assumptions of risk of exposure to wild virus in the US. Major declines in risk of exposure ... could alter the balance significantly' (Hinman *et al.* 1988, p. 295). Despite the high costs of switching from live to killed vaccine, the Advisory Committee on Immunisation Practices recommended the change in 1996 and US vaccine policy delivered killed vaccine exclusively from 2000 onwards (Plotkin and Vidor 2004, p. 1484).

References

Archibugi, D. and Bizzarri, K. (2004) 'Committing to vaccine R&D: a global science policy priority'. *Research Policy* 33(10), pp. 1657–1671

Arnold, A. (2005) 'More than just a vaccine: rising to the challenge of public good provision in a globalising world'. Unpublished MSc dissertation, SPRU, Brighton, University of Sussex

Baldwin, P. (1999) *Contagion and the State in Europe, 1830–1930.* Cambridge, UK: Cambridge University Press

Barder, O. (2005) *Vaccines for Development.* Washington, DC: Center for Global Development

Benison, S. (1967) *Tom Rivers: Reflections on a Life in Medicine and Science.* Cambridge, MA: MIT Press

Blume, S. (1992) *Insight and Industry: On the Dynamics of Technological Change in Medicine.* Cambridge, MA: MIT Press

Blume, S. (2005) 'Lock in, the state and vaccine development: lessons from the history of the polio vaccines'. *Research Policy* 34(2), pp. 159–173

Blume, S. S. and Geesink, I. (2000) 'A brief history of polio vaccines'. *Science* 288(5471), pp. 1593–1594

Bodian, D. (1949) 'Differentiation of types of poliomyelitis viruses'. *American Journal of Hygiene* 49, p. 200

Bodian, D. (1952) 'Experimental studies on passive immunisations against poliomyelitis'. *American Journal of Hygiene* 56, pp. 78–79

Burnet, F. and Jackson, A. (1940) 'The spread of poliomyelitis virus in cynomolgus monkeys'. *Aust J Exp Biol Med* 18(361)

Campbell, D. (1960) 'Blind variation and selective retention in creative thought as in other knowledge processes'. *Psychological Review* 67, pp. 380–400

Carter, R. (1965) *Breakthrough: The Saga of Jonas Salk.* New York: Trident Press

Chase, A. (1982) *Magic Shots: A Human and Scientific Account of the Long and Continuing Struggle to Eradicate Infectious Diseases by Vaccination.* New York: William Morrow

Chataway, J., Brusoni, S., Cacciatori, E., Hanlin, R. and Orsenigo, L. (2007) 'The International AIDS Vaccine Initiative (IAVI) in a changing landscape of vaccine development: a public/private partnership as knowledge broker and integrator'. *European Journal of Development Research* 19(1), pp. 100–117

Constant, E. W. (1980) *Origins of the Turbojet Revolution.* Baltimore, MD: Johns Hopkins University Press

Dosi, G. (1982) 'Technological paradigms and technological trajectories: a suggested interpretation of the determinants and directions of technical change'. *Research Policy* 11(3), pp. 147–162

Dulbecco, R. and Vogt, M. (1954) 'Plaque formation and isolation of pure lines with poliomyelitis viruses'. *Journal of Experimental Medicine* 99, p. 167

Enders, J. (1954) 'Recent advances in the study of poliomyelitis'. *Medicine* 33, pp. 87–95

Galambos, L. and Sewell, J. E. (1995) *Networks of Innovation: Vaccine Development at Merck, Sharp and Dohme, and Mulford, 1895–1995*. Cambridge, UK: Cambridge University Press

Gallagher, H. (1985) *FDR's Splendid Deception*. New York: Dodd, Mead and Company

Garber, D. A., Silvestri, G. and Feinberg, M. B. (2004) 'Prospects for an AIDS vaccine: three big questions, no easy answers'. *Lancet Infectious Diseases* 4(7), pp. 397–413

Hammon, W. (1950) 'Possibilities of specific prevention and treatment of poliomyelitis'. *Pediatrics* 6, pp. 696–705

Hammon, W., Corriel, L., Weherle, P. and Stokes, J. (1953) 'Gamma globulin as a prophylactic agent for poliomyelitis viruses'. *JAMA* 151, p. 1272

Hinman, A. R., Koplan, J. P., Orenstein, W. A., Brink, E. W. and Nkowane, B. M. (1988) 'Live or inactivated poliomyelitis vaccine: an analysis of benefits and risks'. *Am J Public Health* 78(3), pp. 291–295

Hortsmann, D. M. (1952) 'Poliomyelitis virus in blood of orally infected monkeys and chimpanzees'. *Proc Soc Exp Biol Med* 79, pp. 417–419

Hortsmann, D. M. (1985) 'The poliomyelitis story: a scientific hegira'. *Yale Journal of Biology and Medicine* 58, pp. 79–90

Howe, H. A. (1952) 'Antibody response of chimpanzees and human beings to formalin inactivated Trivalent poliomyelitis vaccine'. *American Journal of Hygiene*, 56, pp. 265–279

Klein, A. (1976) Trial by Fury: The polio vaccine controversy. New York: Charles Scribner's Sons

Lanjouw, J. O. (2003) 'Intellectual property and the availability of pharmaceuticals in poor countries'. *Innovation Policy & the Economy* 3(1), pp. 91–129

Lindner, U. and Blume, S. (2006) 'Vaccine innovation and adoption. Polio vaccines in the UK, the Netherlands and West Germany 1955–1965'. *Medical History* (50), pp. 425–446

Longmore, P. K. (1987) 'Uncovering the hidden history of disabled people'. *Reviews in American History* 15, pp. 355–364

Markel, H. (2005) 'Tommy Francis and the Salk Vaccine'. *N Engl J Med* 352(14), pp. 1408–1410

Meldrum, M. (1998) '"A calculated risk": the Salk polio vaccine field trials of 1954'. *BMJ* 317, pp. 1233–1236

Medical Research Council (MRC) (1948) 'Gamma globulin in prevention and attenuation of measles'. *Lancet* 2, pp. 41–44

Mullan, F. (1989) *Plagues and Politics: The Story of the US Public Health Service*. New York: Basic Books

Nathanson, N. and Langmuir, A. (1963) 'The Cutter Incident: Poliomyelitis following formaldehyde inactivated poliovirus vaccination in the US in the Spring of 1955'. *American Journal of Hygiene* 78, pp. 16–81

Nelson, R. R. (2008) 'Factors affecting the power of technological paradigms'. *Ind Corp Change* 17(3), pp. 485–497

Nightingale, P. (2000) 'Economies of scale in experimentation: knowledge and technology in pharmaceutical R&D'. *Ind Corp Change* 9(2), pp. 315–359

Nightingale, P. (2004) 'Technological capabilities, invisible infrastructure and the un-social construction of predictability: the overlooked fixed costs of useful research'. *Research Policy* 33(9), pp. 1259–1284

Offit, P. A. (2005) 'The Cutter Incident, 50 years later'. *N Engl J Med* 352(14), pp. 1411–1412

Oshinsky, D. (2005) *Polio: An American Story*. Oxford, UK: Oxford University Press

Paul, J. (1971) *A History of Poliomyelitis*. New Haven, CT: Yale University Press

Plotkin, S. A. and Vidor, E. (2004) 'Poliovirus vaccine – inactivated' in S. A. Plotkin and W. A. Orenstein (Eds) *Vaccines*. Philadelphia, PA: Saunders, pp. 625–650

Price, D. D. (1984) 'The science/technology relationship, the craft of experimental science, and policy for the improvement of high technology innovation'. *Research Policy* 13(1), pp. 3–20

Racaniello, V. R. (2006) 'One hundred years of poliovirus pathogenesis'. *Virology* 344(1), pp. 9–16

Rinaldo, C. R. Jr (2005) 'Passive immunization against poliomyelitis: the Hammon Gamma Globulin Field Trials, 1951–1953'. *Am J Public Health* 95(5), pp. 790–799

Robbins, F. C. (2004) 'The history of polio vaccine development in S. A. Plotkin and W. A. Orenstein (Eds) *Vaccines*. Philadelphia, PA: Saunders, pp. 17–30

Rogers, N. (1992) *Dirt and Disease: Polio before FDR*. New Brunswick, NJ: Rutgers University Press

Rose, D. W. (2003) *March of Dimes*. Mount Pleasant, SC: Arcadia Publishing

Rosenberg, C. E. (2002) 'The tyranny of diagnosis: specific entities and individual experience'. *Milbank Quarterly* 80(2), p. 237

Rosenberg, N. (1974) 'Science, innovation and economic growth'. *Economic Journal* 84(333), pp. 90–108

Sabin, A. and Olitsky, P. (1936) 'Cultivation of poliomyelitis virus in vitro in human embryonic nervous tissue'. *Proc Soc Exp Biol Med* 31(357)

Salk, J. (1953) 'Studies in human subjects on active immunisation against poliomyelitis'. *JAMA* 151, pp. 1081–1098

Sarewitz, D. (1996) *Frontiers of Illusion: Science, Technology, and the Politics of Progress*. Philadelphia, PA: Temple University Press

Smith, J. S. (1990) *Patenting the Sun*. New York: William Morrow

Thagard, P. (1999) *How Scientists Explain Disease*. Princeton, NJ: Princeton University Press

Time (1953) 'Vaccine for polio'. *Time* magazine

Time (1954) 'Closing in on polio'. *Time* magazine

Tomes, N. (1990) 'The private side of public health: sanitary science, domestic hygiene, and the germ theory, 1870–1900'. *Bulletin of the History of Medicine*, LXIV, pp. 509–539

Vincenti, W. G. (1990) *What Engineers Know and How They Know It: Analytical Studies from Aeronautical History*. Baltimore, MD: Johns Hopkins University Press

Yaqub, O. (2008) 'Knowledge accumulation and vaccine innovation: lessons from polio and HIV/AIDS'. Unpublished DPhil dissertation, SPRU, Brighton, University of Sussex

Yaqub, O. and Nightingale, P. (2012) 'Social science & medicine'. *Social Science & Medicine*, 75(12), pp. 2143–2150

7 Glaucoma

The silent thief of sight

Davide Consoli and Ronnie Ramlogan

Introduction

Glaucoma is the second cause of blindness after cataract worldwide, and the first cause of irreversible vision loss (source: AHAF).[1] Decades of medical scientific research led to the conclusion that 'glaucoma' is most appropriately used with reference to a group of ocular disorders with multi-factorial aetiology (Casson *et al.* 2012). The common feature of them is the degeneration of optic nerve fibres, a neuropathy that leads to blindness if untreated. In contrast with other case studies in this volume, and in spite of continuing efforts on the part of the scientific community, glaucoma's diagnostic and therapeutic solutions are at best partially effective. The onset of disease is often asymptomatic and several forms of glaucoma are simply hard to detect or too aggressive for effective prevention. But, even in the case of timely diagnosis, visual impairment due to glaucoma is still a likely outcome. On the one hand the therapeutic regimes currently available can slow down but not reverse progressive damage to the visual field. On the other hand, glaucoma therapy is still anchored in the tradition of lowering intra-ocular pressure (IOP) in spite of clinical evidence showing that the latter is a risk factor only for some forms of glaucoma. As a result, the efficacy of treatment varies significantly depending on 'which glaucoma' affects an individual, as well as on other factors such as lifestyle and ethnic group.

The persistent shadow of uncertainty that surrounds glaucoma makes it an interesting case for the study of the evolution of medical know-how. The remainder of the chapter offers a synthesis of the key milestones in this area of medical research and practice. Coherent with state-of-the-art glaucoma medicine, the following sections treat advances in scientific understanding and in clinical treatment as separate tracks: this peculiar organisation of the narrative seeks to emphasise the divorce between what is known about disease and what can be done about it.

Glaucoma: a puzzling problem

Using Figures 7.1 and 7.2 as references, let us illustrate the widely accepted classification of the group of diseases that go under the name of glaucoma. There are three main types of disorders that can be further subdivided in sub-categories (Paton and Craig 1976):

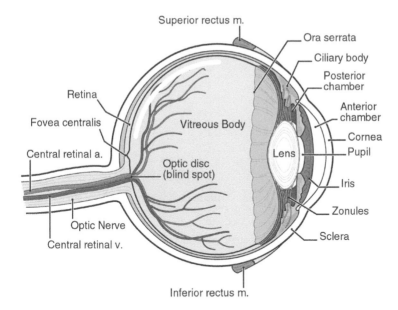

Figure 7.1 Anatomy of the eye.

Source: Health, Medicine and Anatomy Reference Pictures.

(1) Open-angle Glaucoma (OAG) is the most common form of the disease, with an incidence that can vary considerably across countries – from 75 per cent of all glaucoma cases in the United States to much lower percentage shares in Asian countries. Open-angle Glaucoma is due to a particular morphology of the drainage canals between the cornea and the iris (see Fig. 7.1 and the left-hand side of Fig. 7.2) that prevents the outflow of the liquid circulating within the eye and triggers high IOP. In a vast majority of cases the strain due to excess pressure eventually damages the optic nerve in the back of the eye (see Fig. 7.1) and triggers loss of vision. The causes of OAG are mostly unknown and the onset of disease is asymptomatic. There are two variants of this type of glaucoma: Primary OAG has a generally slow onset, about ten years, before damage is detectable; conversely in Normal Tension OAG optic nerve damage is due to inadequate blood supply, so those affected will have normal IOP levels and are generally harder to diagnose. High blood pressure and diabetes are known risk factors for Normal Tension OAG but diagnosis and treatment are at best rudimentary.

(2) Closed-angle Glaucoma (CAG) is due to a congenital feature of the iris (see Fig. 7.1 and right-hand side of Fig. 7.2) that blocks the liquid outflow. Closed-angle Glaucoma accounts for less than 10 per cent of glaucoma cases in the US, but it rises to more than 50 per cent in Asian countries. Contrary to OAG, the development of this form of glaucoma is accompanied by symptoms such

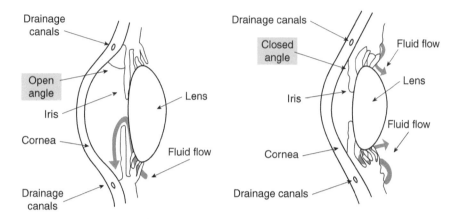

Figure 7.2 Left: fluid dynamics in open-angle glaucoma. Right: closed-angle glaucoma.
Source: Glaucoma.org.

    as severe pain in more than 10 per cent of patients. Closed-angle Glaucoma can be either acute, whereby the closure of the anterior chamber occurs fast and causes rapid IOP increase, or chronic, in which case symptoms are milder and IOP increase gradual. Although the painful symptoms are alert signals, timing is a key factor due to the usually rapid progression of optic nerve damage.

(3) The last group, Developmental Glaucoma (DG), includes rare forms that affect children: 80 per cent of DG cases are diagnosed in the first year of life. The first variant is Primary Congenital Glaucoma, with an incidence of 1 out 10,000 births in the US and accounting for between 50–70 per cent of all cases. The other form, Infantile Glaucoma, is very rare and develops before three years of age.[2] In about ¾ of all cases, DG affects both eyes.

The reader will already appreciate that the shadow of uncertainty that surrounds glaucoma is the crucial point of this story: the majority of patients with the disease do not experience clear symptoms until irreversible visual field loss has already started. Moreover, early diagnosis is difficult for normal tension glaucoma patients because the only recognisable indicator, IOP pressure, is not of use. To make matters more complicated individuals belonging to particular populations with physiologically high IOP levels may never experience optic nerve damage, but the sole presence of abnormal pressure entails periodical check-ups and thus escalating health costs.

    The broader point is that the exceptions to the rule in glaucoma are plenty, and the magnitude of this uncertainty is staggering. In the 2000s the prevalence of glaucoma in the United Kingdom was around 250,000 people, mostly 60 years of age or older, with an estimated misdiagnosis of about 110,000 individuals. In the same period about 50,000 people in Italy were affected by visual impairment due to glaucoma, and an estimated 200,000 were not properly diagnosed. In Germany the incidence is about 23/100,000 cases, mostly among those aged 75 or older. Some 500,000 patients in France receive treatment for POAG, while a broadly similar number of cases remain undiagnosed. Considering direct costs and productivity losses, the economic burden of glaucoma has been estimated at around $2.86 billion every year in the US and US$1.9 billion in Australia. Against this backdrop, the remainder of the chapter will propose a historical journey in the key milestones of this area of medicine.

Glaucoma research: shifting understandings

This section focuses on the long-term development of glaucoma medical research, a story rich in twists, misperceptions and errors that offers an interesting backdrop to appreciate the trial-and-error nature of progress in medicine.

Glaucoma mark one: 1854–1951

The earliest official mention of glaucoma dates back to the medical records of Hippocrates, who is credited with the first systematic classification of abnormal eye appearance in relation to loss of sight.[3] In spite of its ancient roots, Ophthalmology, the branch of medicine that deals with the anatomy, physiology and diseases of the eye, became an independent medical specialty in the mid-nineteenth century (Kansupada and Sassani 1997). The only remaining written records of eye specialists before that time are included within bulky treatises that were so common in the old days of medicine (see, for example, MacKenzie 1833). The clinical assessment and intervention of glaucoma became an independent branch within ophthalmology only in the late 1800s. Indeed, for the best part of the nineteenth century, an examination by the 'eye doctor' consisted of direct observation and evaluation of the iris – the tissue surrounding the pupil that controls light levels inside the eye (use Fig. 7.1 as a reference). As a consequence, abnormalities due to glaucoma, cataract or simple ocular hypertension were ascribed to a single cause and treated accordingly.

As is often the case in medicine, it takes the talents of a pioneer to open up a new direction. Eye surgeon Albrecht von Graefe is credited with having successfully performed the first iridectomy, an incision in the iris of a congested eye. Based on fluid circulation as a model of reference, he conjectured that either excess pressure or hyper-secretion of aqueous humour was the cause of the congestion, and that surgical incision would provide relief by unblocking the flow (Hulke 1859). The experiment marked the first milestone in glaucoma understanding as the blockage of aqueous humour that bathes the lens and the cornea (see Figure 7.1) was described by von Graefe as peculiar and distinct from the other known major eye disease, cataract. Seen from the viewpoint of modern ophthalmology, von Graefe's test was a success for the 'wrong reasons': elevated fluid pressure does indeed obstruct the microcirculation of blood in the eye and, even though the iris is not the locus of aqueous humour production, an incision on the external tissue ends up facilitating the excess outflow. Therefore, contrary to von Graefe's belief, the surgery does not remove the cause of eye congestion, it only alleviates it. As a matter of fact, flow obstruction triggers the degenerative process that impedes the nourishment of optic nerve tissues and ends in loss of sight (see Fig. 7.1).

The logic underpinning von Graefe's iridetcomy was a rudimentary ancestor of the IOP paradigm; a way of understanding glaucoma that will shape, for good or for bad, medical studies and clinical practice for well over a century. Achieving clinical success, notwithstanding limited or incorrect understanding, has been and continues to be a defining feature of problem solving in medicine. Ophthalmology is one among several areas of medical practice where surgeons approached clinical cases in a trial-and-error fashion, with minimal explicit linkages between bedside and bench and no established standards. However tentative by today's standards, von Graefe operated according to a style of practice in which doctors would deduct cause and effect by abstracting from established models in other areas of science such as, for example, the physics of liquid circulation in the case of glaucoma. After von Graefe ophthalmologists started to follow a set of rudimentary guidelines for the diagnosis of glaucoma that included basic tasks such as digital palpation of the eye and a superficial inspection of the pupil and the cornea to detect deviations from 'normal' appearance as recorded in medical books and treatises by professional drawings (Duke Elder 1955).

If, on the therapeutic front, no other option was available other than iris incision, the emergence of new instruments paved the way to different, partially competing, directions in the realm of diagnosis. The catalogue is rich. The tonometer was a device for measuring IOP by recording the resistance of the cornea to indentation (see reviews by Kniestedt *et al.* 2008; Stamper 2011). The associated technique was crucial in the discovery that glaucoma-induced blindness is gradual and not sudden, and that it begins with blind spots in the visual field (scotoma), which are to this day standard early symptoms of loss of vision (see Duke-Elder 1955). Another instrument, the ophthalmoscope, allowed the scanning of the interior of the eye through an atropine-dilated pupil and with the aid of a slit lamp (see Keeler 2003). The enhanced observation of the structural features of the eye led to the detection of the peculiar shape of the diseased optic nerve and the associated excavation of the optic disc. The development of the gonioscope enriched the battery of instruments and techniques to observe the inside of the eye (see Alward 2011). Using this device Barkan established a correlation between glaucoma and various structural features of the eye, such as a particular depth of the anterior chamber and a specific degree of openness of the drainage angle. Finally perimetry stands out from all the 'direct' assessment methods above in that the deterioration of eyesight is gauged directly from the experience of patients (see Johnson 2011). Because of this, the indirect method of perimetry has been for a long time the least preferred method (Trope and Britton 1987).

Consider what each of the above instruments brings to the table. The tonometer provides an objective measure of eye pressure; the ophthalmoscope and the gonioscope rely on the skills of the physician for recognising structural anomalies in the optic nerve; while the perimeter entails the skills of the patients in describing the extent of visual impairment. It is not surprising that the style of practice of glaucoma specialists evolved towards the integrated use of all these instruments (Duke Elder 1955). Unlike other areas of medicine where the margin of ambiguity is smaller, no single device was sufficient to become a 'gold standard' on its own.

Furthermore, once they became part of the standard practice these techniques triggered further, at times unintended, discoveries. In the case of the ophthalmoscope, the discovery of cupping brought to the fore limitations of the existing clinical procedures (Albert and Edwards 1996). If indeed the anomalous disc cupping played a role in the onset of glaucoma, 'how much' cupping was to be considered clinically dangerous? Also, what is the correlation between damage in the optic nerve and loss of visual field? Further down the line, attempts to address these issues would lead to discoveries that undermined the very framework that originated them, namely the IOP paradigm. The diffusion of the gonioscope led to similarly revolutionary, and unintended, consequences. Until then intra-ocular liquid blockage was understood as an anomaly due to the narrow angle in the anterior chamber of the eye, but repeated observations through a gonioscope allowed Barkan to observe for the first time a central feature of glaucoma: the open angle (Lowe 1995). This breakthrough would lead to a radical new understanding of disease and the differentiation of different forms of eye disorders that go under the umbrella name of glaucoma. Even perimetry, often dismissed because it is an indirect assessment method, has found its proper space in the changing practice of glaucoma ophthalmology, especially for early stage glaucoma when the incidence of visual loss is small and difficult to identify. The clinical diffusion of perimetry led to new standards of interaction between patients and eye clinicians with fine-tuning in the ability to communicate and collaborate.

The first International Symposium on glaucoma in 1952 marks the end of this early phase in the history of glaucoma research. By bringing together practitioners from all over the world, this event provided identity to a hitherto dispersed scientific community and promoted glaucoma to an independent sub-specialty within the broad realm of ophthalmology. The symposium represents a noticeable point of discontinuity for the system of thinking of the ophthalmologic community because the articulated summary of different disease manifestations was tangible proof that the extent of its diversity was greater than had hitherto been appreciated. One of the key outcomes was the first international classification of glaucoma, based on the distinction between Open-Angle and Closed-Angle, obtained by systematically collating and comparing evidence from different sources (Duke-Elder 1955).

Glaucoma mark two: 1952–1990s

Until the early 1950s, both in America and elsewhere, ophthalmology was taught as a part of the basic medical curriculum due to overall low prospects of specialising in what was considered an incurable disease (Liesegang *et al.* 2003). Despite the American Ophthalmological Society being operative since 1864, it was, in fact, the first medical specialty organisation in the US; no specialised scientific outlet existed besides the Acta of the society's meetings (Newell 1997). However, as recounted above, the absence of a proper scientific infrastructure was no obstacle to individual practitioners who developed the array of scientific instruments outlined, tools that also shaped complementary skills such as criteria for judging eye appearance and for performing surgical incisions. In the years to come, consolidation of clinical modalities would disclose further remarkable differences in symptomatology and incidence across forms of glaucoma. This evidence would usher in a new era, building on the acknowledgement that the damage due to the degenerative process of the optic nerve cannot be reversed but that it can be slowed down. As a consequence, the central ethos of clinical glaucoma became 'managing the disease' as opposed to removing its causes altogether. This had significant consequences for the criteria underpinning the design of both clinical practice and medical training, as synthesised in Table 7.1. The new understanding of glaucoma as a family of eye conditions adds to the above, and leads to the partitioning of a meta-problem into various sub-domains according to the established typologies and the associated modalities of research and practice.

At a more fundamental level the newly accepted heterogeneity of the disease raised doubts about the foundations of the IOP paradigm. This is confirmed by a stream of epidemiological studies aimed at testing the (hitherto accepted) co-occurrence between ocular pressure and glaucoma. Following the pioneering steps of Hollows and Graham (1966) such studies demonstrated that IOP is only a risk factor, and that only about 10 per cent of people with increased abnormal pressure are affected, while about 25 per cent of glaucoma patients have normal pressure levels (see Liesegang 1996). These groundbreaking discoveries added substantive uncertainty concerning the aetiology of the disease and caused a stall in medical scientific thought throughout the 1960s, and the glaucoma medical specialty only saw marginal improvements in the existing therapeutic and diagnostic regimes of IOP measurement and reduction.

This status quo changed in the mid-1970s when the spur of new ideas and ever more specialised conjectures inspired by clinical, epidemiologic and laboratory research ushered in a new era (see Table 7.1). The accepted diversity of the disease prompted efforts aimed at perfecting the classification of glaucoma, thus paving the way to a variety of research directions. One approach set out to explore associations between glaucoma and other diseases, especially diabetes and heart conditions, in search of regularities that could elucidate on aetiology. Another important branch of research stemmed from the rediscovery of intuitions that had been recorded back in the old days but that remained overseen or not understood properly; most prominent among these was the excavation of the optic nerve known as 'cupping', which became the focus of much attention and the key to fundamental developments between the mid-1960s and the late 1970s. As the incidence of IOP was downsized as a risk factor, practitioners who also engaged in research focused on the linkages between eye pressure and damage in the optic nerve. This meta-hypothesis was articulated in two main directions: the mechanical theory had it that intraocular pressure exerts a force that compresses

Table 7.1 Clinical and scientific steps in the evolution of glaucoma

	1854–1951	1952–1970	1971–1990	1991–2003
Meta-hypothesis	Intra-ocular pressure	Different types of glaucoma Primary Open Angle Glaucoma Angle Closure Glaucoma Normal Tension Glaucoma Acute Glaucoma Child Glaucoma	Progressive glaucomatous damage Neuropathy - Vascular cause - Mechanical cause Hereditary components Epidemiology studies Association with other diseases	Optic nerve fibres erosion Cellular anomaly Genetic mutations
Diagnosis	Measurement IOP > Tonometer Observation Eye fundus > Ophthalmoscope Iris angle > gonioscope	Qualitative functionality Assessment Visual field > perimeter	Digital tonometer Laser ophthalmoscope Computerised perimeter Qualitative structure assessment Optic nerve > digital photography	Quantitative structure assessment: Retinal nerve fibre layer > Scanning laser tomography Scanning polarimeter Optical coherence tomography
Therapy	IOP-lowering eye incision: iridectomy	Trabulectomy Drainage	Laser iridectomy Laser trabulectomy IOP lowering drugs: Beta-Blockers	Combinations of eye drops Neurochemistry
Division of labour **Skills**	Isolated practitioners Surgical Eye palpation Direct observation	Intra-organisational collaboration University + hospital	Hospital + firms; university + firms Cross-disciplinary interaction Indirect observation	Team-working Patient management Data handling

Source: authors' own elaboration.

the optic nerve thus altering its functionality; the vascular theory conversely posited that high pressure damages the optic nerve by reducing the nourishment from the blood supply. A series of important breakthroughs by Drance and Begg (1970) and Begg *et al.* (1971) laid the foundations of glaucoma as a neuropathy. In the following decade the path-breaking work by Airaksinen and Tuulonen (1984) finally refuted the idea that optic disc changes are always a factor in glaucoma, and highlighted that the pathogenesis of the disease differs to a substantial degree. Finally, Høvding and Aasved (1986) established the impact of family history on glaucoma patients. It is now accepted that neither theory explains how optic nerve damage occurs across different types of glaucoma, and that the effects described by either probably work in combination rather than being mutually exclusive (Geijssen 1991).

The proliferation of hypotheses concerning the aetiology coincides with the emergence of new instrumentation and parallel advances in digital imaging, laser and ultrasound techniques. In the second half of the 1970s clinical researchers unveiled a correlation between the erosion of the optic nerve in glaucoma and visual field loss that contributed enormously to shift the reference model from hydraulic blockage to neuropathy (Drance 1975). Accordingly IOP-based assessment progressively made room for alternative techniques such as analysis of eye structure (optic nerve) and of its functionality (visual field). This was aided at first by the adoption of digital photography for qualitative assessment of the optic nerve and, subsequently, by the development of the scanning laser ophthalmoscope for quantitative assessment (Quigley 1998; Sharp *et al.* 2004). At a more fundamental level, the embracing of indirect observation mediated by the technological devices underscores the evolving logic of diagnostics since glaucoma was no longer understood as a 'binary' but rather as a progressive disease in which chances of detection and of effective treatment depended on the stage of the neuropathy, and in which the boundaries of clinical practice drifted in pursuit of the 'mechanical objectivity' (Daston and Gallison 2007) promised by visual techniques. In this eventful phase the skill base of glaucoma specialists expanded to facilitate interaction with formerly unrelated areas of science; as Consoli and Ramlogan (2008) show, the universe of scholarly research on glaucoma mushroomed in journals on neurology, cellular biology and genetics after the mid-1990s.

Glaucoma mark three: 1990s-2000s

During the 1990s glaucoma research reaches out to genetics, cellular biology and molecular science. At the onset of this was the discovery that incidence varies significantly according to age, family history and, most cogently, racial background. As Figure 7.3 shows, the likelihood of blindness or visual impairment due to glaucoma is significantly higher for African Americans in the US than for Caucasians (source: National Eye Institute, US).[4] Likewise, the prevalence is significantly higher in Africa and Japan when compared with the rest of the world (Fig. 7.4).

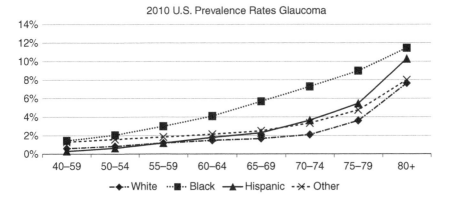

Figure 7.3 2010 US prevalence rates for glaucoma.

Source: National Eye Institute.

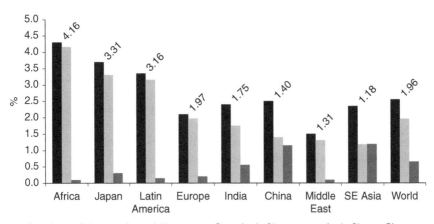

Figure 7.4 Glaucoma prevalence by type and country.

Source: Kyari *et al.* 2013.

On the other hand, the finding that glaucoma tends to run in families laid the ground for the discovery of the gene responsible for Primary Open Angle Glaucoma in the 1990s (Stone *et al.* 1997). However, initial enthusiasm dwindled as soon as it became clear that only three per cent of cases are associated with defects in single genes, and that no single mode of inheritance adequately describes the disease. It seems that glaucoma depends not so much on a single gene but rather upon the interaction of several in combination with environmental factors (Wiggs 2007). This is to say that despite remarkable progress, translational gaps hold at bay improvements in practice and that the scenario of routine genetic screening remains remote.

Another important thread in glaucoma research follows on the path set by neuropathy studies back in the mid-1970s, in particular the discovery of erosion in the optic nerve fibres due to glaucomatous damage (Hoyt *et al.* 1973). The emergent diagnostic modality has progressively shifted from direct observation of the optic nerve to computerised assessments of the thickness of the optic nerve fibres (Retinal Nerve Fibre Layer, RFNL) (Sommer *et al.* 1984). In this new regime technologies such as Scanning Laser Polarimetry (by Laser Diagnostic Technologies Inc.), Laser Retina Tomography (by Heidelberg Engineering Inc.) and the Optical Coherence Tomography based on ultrasound (Zeiss-Meditec Inc.) are used to produce quantitative measures of optic disc structure and nerve consistency, and to calculate indices for monitoring morphologic change due to disease progression over time (Schuman and Lemij 2000). It is worth stressing that these forms of computerised assessments carry both benefits as well as problems. Some of these are technical: for example, computerised data collection makes sense within a regime of longitudinal testing to detect disease progression in individual patients, but this particular task requires data handling skills from humans as well as compatibility across data collected at different points in time and, given the rapid pace of technical advance, by presumably different devices (Trick *et al.* 2006). Parallel to this stand concerns among ophthalmologists who seek to reaffirm the value of 'trained judgment' (Daston and Gallison 2007) as the dominant scientific style in which sophisticated artefacts capture but do not discern meaningful patterns.[5]

The preceding observations underscore significant transformations in the skill base of glaucoma specialists, whose modern version is far removed from the eye surgeons of a century ago. Glaucoma specialists work in teams where traditional medical skills blend with interpersonal communication skills to engage both co-workers (ophthalmic nurses, optometrists, clinic managers, etc) and patients. Since age is a significant risk factor for ocular pressure, ageing populations account for a progressively higher demand for eye treatment, as well as for long-term disease management skills as opposed to one-off interventions. Empirical work on the dynamics of scientific networks confirms this by showing the 'broadening' of the ecology of journals in which glaucoma papers are published, which now include areas as diverse as pharmacogenomics, biological chemistry, molecular genetics and so forth (Ramlogan and Consoli 2013). Similarly the institutional and geographic distributions of glaucoma research efforts have evolved remarkably. Consoli and Ramlogan (2012) find that after the 1970s Europe's contribution to glaucoma research has increased on a par with North America – together they account for more than 70 per cent of all published work. Concerning the typology of organisations involved in research, the bulk of effort is located in mixed institutions like university hospitals, with a very scant role played by firms. This is not surprising considering the state-of-the-art of medical scientific research on glaucoma and the fact that, besides a few diagnostic devices, market opportunities of the kind that typically attract firms have been scarce. On the whole the geography of collaboration and the division of labour across the scientific community has changed as a response to persistent uncertainty about the nature of the disease.

Glaucoma therapeutics: the persistence of old ways

As already anticipated in the preceding pages, progress in glaucoma therapeutics did not go as far as scientific understanding. Established pharmaceutical and surgical therapies have limited capacity to stop glaucoma progression, arguably because they are designed according to the logic of reducing IOP, a risk factor but not necessarily the sole cause of disease. In any case, low IOP does not protect *de ipso facto* from glaucoma progression due to the influence of other risk factors (Sommer *et al.* 1984). At the same time the exact role of these factors is not sufficiently understood so as to provide parameters and criteria for either therapeutic or preventive treatments. This lack of understanding limits glaucoma therapy to various forms of IOP long-term management, while a wide margin of clinical uncertainty persists even in the most established areas of intervention. Indeed, while it is known that keeping IOP low has beneficial effects in slowing down the progression of visual loss and in preserving the visual function, it has so far proven impossible to establish 'optimal target levels' of liquid pressure. It is also important to reiterate that there are different forms of glaucoma, and that no single solution qualifies as gold standard. Each patient is treated according to a particular regimen that may involve drug treatment, laser or filtering surgery, or a combination of both. The choice of therapy depends on specific characteristics such as eye structure, age, stage of glaucoma, and other risk factors. Let us illustrate the main features of the existing approaches.

Eye surgery

Modern eye surgery techniques can be broadly divided into conventional- and laser-based procedures. The former consists of filtering operations aimed at creating a drainage channel in the inner layers of the eye (from the anterior chamber to the external surface of the eye – see Fig. 7.1 for reference). Recall that von Graefe performed the first recorded eye incision (iridetcomy) in the mid-1800s, followed in 1867 by the first external filtration procedure to drain fluid outside the eye. Those experiences provided the basis for glaucoma surgery, and future advances in materials, filtering techniques and drainage devices would stay coherent to the IOP reducing regime (Razeghinejad and Spaeth 2011). A popular type of surgery is trabeculectomy, a technique that became the gold standard for primary open- and closed-angle glaucoma in the late 1960s (Cairns 1968). These days surgeons see trabeculectomy as a 'last resort' when prior pharmacological treatment has failed to reduce IOP (AAO 2005). Indeed, while it can successfully reduce pressure and prevent loss of vision in standard glaucoma cases, trabeculectomy rarely suffices and patients often have to undergo further surgery or other treatments. Moreover, due to significant differences in the onset of the disease across patient groups, trabeculectomy is known to be less successful in African Americans, in children affected by congenital glaucoma, in people affected by neo-vascular secondary glaucoma (a rare form in which new blood vessels grow in the eye), and in people who had previous eye surgery. In sum, trabeculectomy is a palliative for some types of glaucoma but is not a universal solution.

The second regime of surgical intervention is based on the use of a laser to produce a very small burn, or opening, in the eye and to increase the flow of fluid. Since its adoption in the late 1970s (Wise and Witter 1979) laser trabeculoplasty (LTP) in particular has been modified to improve results and reduce complications, such as a postoperative increase of IOP. Laser eye surgery is a matter of routine these days and can be performed in an outpatient setting. After initial success LTP proved less than satisfactory for long-term control of glaucoma. The most recent available statistics from the US show that laser trabeculoplasty continues to be the dominant surgical procedure, more so than trabeculectomy (see Fig. 7.5; source: Corcoran 2009) and other surgical procedures.

Alternative techniques such as Selective Laser Trabeculoplasty (SLT) or Micropulse Laser Trabeculoplasty (MLT) have been known to be more effective, simpler to perform and more cost effective, though the evidence in support of their higher efficacy is scant as yet. In any case laser surgery has benefits that are only temporary, it works only for Open-Angle Glaucoma, the most common form of disease, and is not a long-term solution (Razeghinejad and Spaeth 2011).

Glaucoma drugs

Modern glaucoma pharmacology begins around the 1860s, when understanding of the disease was still vague but when eye doctors would experiment

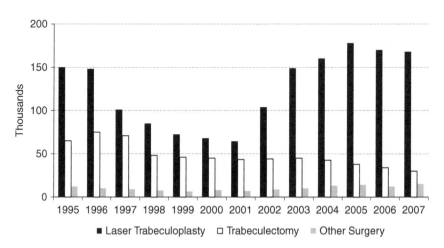

Figure 7.5 Medicare claims for glaucoma surgery.

Source: Authors' own elaboration from Corcoran 2009.

with natural remedies on patients presenting a swollen eyeball, then the only recognisable symptom of glaucoma. English pharmacologist and physician Sir Thomas Fraser is credited with having heralded the era of glaucoma pharmacology with a publication on the *Physiological Action of the Calabar Bean* (sic) in 1867. Fraser's interest lay not in glaucoma but rather in further understanding of the properties of a tropical Africa bean that was known to be poisonous if ingested but that, he conjectured, also had a beneficial effect on nerves and tissues. He focused on the eye due to the ease of observation of changes in the condition of the pupil, and observed that after injection the tissues would relax after an initial contraction (Fraser 1867). The experiment was successful in inducing decongestion of the eyes of both animals and humans and marked the appearance of the first IOP-lowering medication (Realini 2011). Interestingly, the calabar bean was also used by von Graefe during the famous iridetcomy experiment, but as a miotic – that is, as a substance that induces tissue constriction: in other words, the eye surgeon appreciated only some of the properties of the bean while ignoring the relaxing effects.

The demonstrated effectiveness of surgery on the limited (with hindsight) portion of glaucoma cases that could be diagnosed by early 1900 know-how, did not encourage further explorations on the eye drug front. New topical eye drops were adopted in ophthalmology more as a consequence of serendipitous discoveries than as part of an intended therapeutic strategy. This was the case with substances that would later be important in the development of glaucoma pharmacology such as Epinephrine, a hormone that is used as neurotransmitter, and Acetazolamide, a carbonic anhydrase inhibitor with fluid retention properties. To illustrate, Epinephrine, the first medication for glaucoma, became commercially available only in the 1950s (Realini 2011). Likewise, despite proven IOP-lowering properties, carbonic anhydrase inhibitors were initially used only for antibacterial and diuretic applications because of the side effects. This discouraged the development of similar topical drugs until the 1970s when industry-university synergy between Merck and the University of Florida led to the discovery of Dorzolamide, the base for the first successful topical carbonic anhydrase inhibitor (Zimmerman 1977) that is still available on the market today. This discovery stands out as the first systematic search for a workable molecule for glaucoma (Realini 2011).

On the back of newly discovered characteristics of the disease, reducing excess pressure was viewed as a preventive means to slow the deterioration of the optic nerve. The new pharmacological regime further accelerated in the 1970s after the appearance of beta-blockers, a class of drugs that became the medical therapy of choice (Rafuse 2003). The glaucoma drugs currently available on the market can be divided into two broad classes depending on the type of action:

1. Beta-blockers, adrenaline drops (sympathomimetics) and carbonic anhydrase inhibitors reduce the amount of fluid being secreted into the eye;
2. Prostaglandin analogues and miotics (parasympathomimetics) instead increase the drainage of fluid out of the eye.

Recent survey data indicate that prostaglandin analogues are the preferred option for first-line treatment while beta-blockers are more commonly prescribed for second-line treatment (Nasser and Stewart 2006). Moreover physicians have a preference for fixed combinations in glaucoma therapy because of perceived improved compliance, patient convenience and reduced toxicity from preservatives.

Almost paradoxically, the timing of progress in the pharmacological regime coincides with the recognition that IOP-lowering is irrelevant for about half of the known forms of glaucoma, and has short-lived efficacy where it applies. Probably as a result of this, there has been very little exploration in new directions over the last 30 years, and the expansion of existing treatments only responds to the need for minimising side effects for patients. The only recent advances worth mentioning are the development of new delivery systems for drugs. Recall that glaucoma is a chronic disease, mostly symptomless, that affects relatively more elderly people: this mix of characteristics is believed to be the main culprit for poor compliance with prescriptions. Recent estimates suggest that as much as 60 per cent of patients in the US fail to maintain the medication regimen (Rossi *et al.* 2011). A response to this problem is the use of specific devices that facilitate drug release. Some delivery vehicles are depot forms, others are variants of contact lenses designed to reside in the eye to deliver drugs in a continuous, slow-release fashion (Abelson *et al.* 2013).

Changing patterns in glaucoma therapy

The appearance of effective topical drugs in the mid-1990s shifted glaucoma therapeutics towards increasing reliance on drugs to the detriment of surgery, both laser and traditional procedures. Figure 7.6 shows this pattern in the US (source: Corcoran 2009) but recent research confirms this has been the case across a number of countries: Ireland (Fraser and Wormald 2008), England (Keane *et al.* 2008), Scotland (Bateman *et al.* 2001), Canada (Campbell *et al.* 2008), Australia (Walland 2004), the Netherlands (van der Valk *et al.* 2005) and Italy (Ricci *et al.* 2008).

As the reader will have appreciated by now, existing therapy is effective if (i) glaucoma can be diagnosed in a timely fashion and (ii) it is a form of glaucoma for which surgical and pharmacological IOP-lowering regimes work. These two preconditions leave out large chunks of the population. To illustrate, it is estimated that over the last decade more than four million people affected by glaucoma in Europe were undiagnosed and untreated. On a global scale, it is estimated that up to 50 per cent of those with glaucoma in the industrialised countries do not receive proper treatment until it is too late (Sommer *et al.* 1991). Against this backdrop, it is not surprising that medical research is continuously trying to reduce the margin of uncertainty by moving diagnostics and therapy beyond the IOP paradigm. The most promising avenue is the search for direct approaches to treat and prevent glaucomatous neuropathology following up the recent discovery that some IOP-reducing drugs already in use, namely glutamate antagonist,

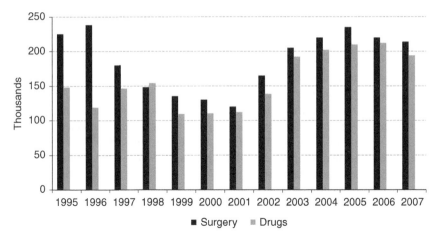

Figure 7.6 Total glaucoma procedure claims by type in the US.

Source: authors' own elaboration from Corcoran 2009.

also have neuro-protective effects. Yet further encouraging results come from the discovery of strong similarities between glaucoma neuropathy and the neuro-degeneration that is commonly at the root of Alzheimer's disease (Kent 2007). But both the latter avenues are still in their infancy and, in the absence of profound breakthrough, the frontier of glaucoma therapy is likely to be based on the combined use of IOP-lowering drugs and generic neuro-protectors that have been recently integrated in the armamentarium more as a result of a serendipitous discovery than of intended strategy. It looks like the goal of preserving visual function in patients with glaucoma is not yet on the horizon.

Concluding remarks

This chapter has illustrated the persistent uncertainty that characterises glaucoma. Different types of eye disease go under that name, and medical understanding on its aetiology and pathogenesis is ridden with imprecision as much as clinical practice is obstructed by significant limitations. Patients with normal pressures can show visual field loss, while the visual fields in others with high IOPs may remain intact. The consequences are staggering both in terms of incidence and of health costs.

The history of this disease is a prime example of how the efficacy of scientific conjectures and of diagnostic and therapeutic avenues has been severely biased by incorrect framing and the persistence of the latter in the clinical realm. It is worth concluding with some reflections on what this analysis adds to the existing body of work on medical innovation. In this sense the present chapter reaffirms the dynamicity and uncertainty of the discovery process that enables improvements in diagnosing, healing and preventing human sickness. Transliterating Simon (1969), the search for solutions to medical problems is beset by substantive ignorance concerning the cause of disease as well as by procedural uncertainty about how to tackle it. Progress in life sciences has gone through different regimes for organising healthcare. At the heart of this process are the removal of cross-disciplinary barriers (Rosenberg 2009) and the intensification of experimental regimes for manipulating, reproducing and evaluating physiological phenomena (Gelijns *et al.* 2001; Nelson *et al.* 2011). The correlation of these two forces, horizontal dynamism across disciplinary boundaries and vertical transmissions between basic scientific and experiential realms, has provided a major step change in the production, diffusion and use of knowledge in medicine (Consoli and Mina 2009). The history of glaucoma shows that the major breakthrough, the shift from hydraulic blockage to neurological paradigm, stemmed from the latter force, systematic organisation of experiential observation, and developed further, guided by the former, the abridging of ophthalmology with other specialties that had much to offer.

The lack of agreed clinical solutions acting as a focusing device – like stents for the treatment of coronary artery disease or intra-ocular lens for cataracts – places glaucoma far from the medical areas that have been studied so far by innovation scholars, as the understanding of the disease is limited, indications for prevention are scarce, the margin of error for diagnostics is still significant and the therapeutic strategies are backward with respect to what glaucoma is not – i.e. an exclusively IOP-related condition. But this does not make it a less interesting story. Perhaps, as Rosenberg (1976, p. 278) claimed, success stories tend to capture our attention more than unsuccessful ventures, but diehard problems disclose equally important aspects of what problem solving and the pathways through the search space look like when the problem remains ill defined. To be sure, the persistence of an inaccurate paradigm is a forceful reminder of the extent to which 'knowledge permits' but 'also constrains' (Rosenberg 1976, p. 278). Let us conclude with a final remark on the way ahead for the emerging strand of research on medical innovation: we are convinced that the continuing addition of empirical works is not merely feeding an inventory but that, as suggested above and elsewhere, the pathways of learning in the universe of medicine unfold and combine in different ways and along trajectories that we are just beginning to understand. Let the journey continue.

Notes

1 American Health Assistance Foundation. Available online at: http://www.ahaf.org/glaucoma/about (last accessed 30 April 2010).
2 Source: Glaucoma Research Foundation. Available online at: http://www.glaucoma.org/uploads/grf_childhood_glaucoma.pdf
3 In fact, the disease is named after the ancient Greek γλαυκος evoking the greenish-blue or sea-green colour of the glaucomatous eye.
4 See The Salisbury Eye Evaluation Study, *Arch Ophthalmol* 2000.
5 See, for example, Weinreb (2003, p. 201) 'With structural tests, we have to recognize that there is no perfect instrument. There wasn't one in 1994, there isn't one in 2003, and there won't be one in 2010. Every instrument, whether objective or subjective, has advantages and disadvantages'.

References

AGIS Investigators, The Advanced Glaucoma Intervention Study (AGIS) (2000) 'The relationship between control of IOP and visual field deterioration'. *American Journal of Ophthalmology*, 130, pp. 429–440

Abelson, M. B., David, R., Shapiro, A. and McLaughlin, J. (2013) 'A look at the safety and efficacy of various therapies for the disease, from pharmaceuticals to surgical implants'. *The Review of Ophthalmology* 13(3)

Airaksinen, P. J. and Tuulonen, A. (1984) 'Early glaucoma changes in patients with and without an optic disc haemorrhage'. *Acta Ophthalmolica* (Copenh), Vol. 62, No. 2, pp. 197–202

Albert, D. M. and Edwards, D. D. (1996) *The History of Ophthalmology: Glaucoma* (1st Ed.). Oxford, UK: Blackwell Science

Alward, W. L. M. (2011) 'A history of gonioscopy'. *Optometry & Vision Science*, 88(1), pp. 29–35

American Academy of Ophthalmology (AAO) (2005) *Primary Open-Angle Glaucoma, Limited Revision (Preferred Practice Pattern)*. San Francisco: American Academy of Ophthalmology

Bateman, D. N., Clark, R., Azuara-Blanco, A., Bain, M. and Forrest, J. (2001) 'The impact of new drugs on management of glaucoma in Scotland: observational study'. *British Medical Journal*, 323, pp. 1401–1402

Begg, I. S., Drance, S. M. and Sweeney, V. P. (1971) 'Ischaemic optic neuropathy in chronic simple glaucoma'. *British Journal of Ophthalmology*, Vol. 55, No. 2, pp. 73–90

Cairns, J. E. (1968) 'Trabeculectomy. Preliminary report of a new method'. *American Journal of Ophthalmology*, 5, pp. 673–9

Campbell, R. J., Trope, G. E., Rachmiel, R. and Buys, Y. M. (2008) 'Glaucoma laser and surgical procedure rates in Canada: a long-term profile'. *Canadian Journal of Ophthalmology/Journal Canadien d'Ophtalmologie*, 43(4), pp. 449–453

Casson, R. J., Chidlow, G., Wood, J. P., Crowston, J. G. and Goldberg, I. (2012) 'Definition of glaucoma: clinical and experimental concepts'. *Clin Exp Ophthalmol* 40 (4), pp. 341–9

Cerulli, L., Cedrone, C., Cesareo, M. and Palma, S. (1997) 'L'epidemiologia del glaucoma' in L. Cerulli, M. Miglior and F. Ponte (Eds) *Epidemiologia in Italia*. Rome: Relazione ufficiale al LXXVII Congresso della Società Oftalmologica Italiana, pp. 163–246

Consoli, D. and Mina, A. (2009) 'An evolutionary perspective on health innovation systems'. *Journal of Evolutionary Economics*, 19(2), pp. 297–319

Consoli, D. and Ramlogan, R. (2008) 'Out of sight: problem sequences and epistemic boundaries of medical know-how on glaucoma'. *Journal of Evolutionary Economics*, 18(1), pp. 31–56

Consoli, D. and Ramlogan, R. (2012) 'Patterns of organization in the development of medical know-how: the case of glaucoma research'. *Industrial and Corporate Change*, 21(2), pp. 315–343

Corcoran, K. J. (2009) 'Shifting tides of glaucoma surgery'. *Review of Ophthalmology* 9(36)

Daston, L. and Galison, P. (2007) *Objectivity*. New York: Zone Books

Drance, S. M. (1975) Correlation of optic nerve and visual field defects in simple glaucoma. *Transactions of the Ophthalmologic Societies of the United Kingdom*, 95(2), pp. 288–296

Drance, S. M. and Begg, I. S. (1970) 'Sector haemorrhage: a probable acute ischaemic disc change in chronic simple glaucoma'. *Canadian Journal of Ophthalmology*, Vol. 5 No. 2, pp. 137–141

Duke-Elder, S. (Ed.) (1955) *Glaucoma: C.I.O.M.S. Symposium*. Oxford, UK: Blackwell

Fraser, S. G. and Wormald, R. P. L. (2008) 'Hospital episode statistics and changing trends in glaucoma surgery'. *Eye*, 22, pp. 3–7

Fraser, T. R. (1867) 'On the physiological action of the calabar-bean (*Physostigma venenosum Balf.*)'. *Transactions of the Royal Society of Edinburgh* XXIV

Fraunfelder, F.T. and Meyer, S. M. (1989) 'Systemic adverse reactions in glaucoma medications'. *Internal Ophthalmology Clinician*, 29, pp. 143–6

Geijssen, H. C. (1991) *Studies on Normal Pressure Glaucoma*. Amsterdam: Kugler Publications

Gelijns, A., Graff Zivin, J. and Nelson, R. R. (2001) 'Uncertainty and technological change in medicine'. *Journal of Health Politics, Policy, and Law*, 26, pp. 913–924

Glaucoma Laser Trial Research Group (1995) 'The Glaucoma Laser Trial (GLT) and Glaucoma Laser Trial follow-up study: 7. Results'. *American Journal of Ophthalmology*, 120, pp. 718–31

Glick, H., Brainsky, A., McDonald, R. C. and Javitt, J. C. (1994) 'The cost of glaucoma in the United States in 1988'. *Chibret Int J Ophthalmol.*, 10, pp. 6–12

Grant, W. M. and Burke, J. F. Jr (1982) 'Why do some people go blind from glaucoma?' *Ophthalmology*, 89, pp. 991–8

Gray, S. F., Spry, P. G., Brookes, S.T., Peters, T. J., Spencer, I. C., Baker, I. A., Sparrow, J. M. and Easty, D. L. (2000) 'The Bristol shared care glaucoma study: outcome at follow up at 2 years'. *Br J Ophthalmol.*, 84, pp. 456–63

Heijl, A., Leske, M. C., Bengtsson, B., Hyman, L., Bengtsson, B. and Hussein, M. (2002) 'Reduction of intraocular pressure and glaucoma progression: results from the Early Manifest Glaucoma Trial'. *Arch Ophthalmology*, 12, pp. 1268–79

Høvding, G. and Aasved, H. (1986) 'Prognostic factors in the development of manifest open angle glaucoma. A long-term follow-up study of hypertensive and normotensive eyes'. *Acta Ophthalmologica* (Copenh), Vol. 64, No. 6, pp. 601–608

Hoyt, W. F., Frisen, L. and Newman, N. M. (1973) 'Fundoscopy of nerve fiber layer defects in glaucoma'. *Investigative Ophthalmology and Visual Science*, 12, pp. 814–829

Hulke, J. W. (1859) 'Iridectomy in glaucoma'. *Lancet*, November 19, p. 515

Iskedjian, M., Walker, J., Vicente, C., Trope, G. E., Buys, Y., Einarson, T. R. and Covert, D. (2003) 'Cost of glaucoma in Canada: analyses based on visual field and physician's assessment'. *J. Glaucoma*, 12, pp. 456–62

Johnson, C. A. (2011) 'A history of perimetry and visual field testing'. *Optometry & Vision Science* 88(1), pp. E8–E15

Johnson, Z. K., Griffiths, P. G. and Birch, M. K. (2003) 'Nurse prescribing in glaucoma'. *Eye*, 17, pp. 47–52

Kansupada, K. B. and Sassani, J. W. (1997) 'Sushruta: the father of Indian surgery and ophthalmology'. *Documenta ophthalmologica*, 93(1–2), pp. 159–167

Keane, P. A., Khan, M. I., Saeed, A., Stack, J., Tormey, P., Hayes, P. and Beatty, S. (2008) 'The impact of available anti-glaucoma therapy on the volume and age profile of patients undergoing glaucoma filtration surgery'. *Eye*, 23(8), pp. 1675–1680

Keeler, C. K. (2003) 'A brief history of the ophthalmoscope'. *Optometry in Practice*, 4, pp. 137–145

Kent, C. (2007) 'Glaucoma drugs: the search for new options'. *Review of Ophthalmology* 7(10)

Kniestedt, C., Punjabi, O., Lin, S. and Stamper, R. L. (2008) 'Tonometry through the ages'. *Survey of Ophthalmology*, 53(6), pp. 568–591

Kobelt, G. and Jonsson, L. (1999) 'Modelling cost of treatment with new topical treatments for glaucoma. Results from France and the United Kingdom'. *International Journal of Technology Assessment in Health Care*, 15, pp. 207–19

Kobelt-Nguyen, G., Gerdtham, U. G. and Alm, A. (1998) 'Costs of treating primary open-angle glaucoma and ocular hypertension: a retrospective, observational two-year chart review of newly diagnosed patients in Sweden and the United States'. *Journal of Glaucoma*, 7, pp. 95–104

Krumpaszky, H. G., Lüdtke, R., Mickler, A., Klauss, V. and Selbmann, H. K. (1999) 'Blindness incidence in Germany. A population-based study from Wurttemberg-Hohenzollern'. *Ophthalmologica*, 213, pp. 176–82

Kyari, F., Abdull, M. M., Bastawrous, A., Gilbert, C. E. and Faal, H. (2013) 'Epidemiology of glaucoma in Sub-Saharan Africa: prevalence, incidence and risk factors'. *Middle East Africa Journal of Ophthalmology*, 20(2), pp. 111–125

Lichter, P. R., Musch, D. C., Gillespie, B. W., Guire, K. E., Janz, N. K., Wren, P. A. and Mills, R. P. (2001) 'Interim clinical outcomes in the Collaborative Initial Glaucoma Treatment Study comparing initial treatment randomized to medications or surgery'. *Ophthalmology*, 108, pp. 1943–53

Liesegang, T. J. (1996) 'Glaucoma: changing concepts and future directions'. *Mayo Clinical Proceedings*, 71, pp. 689–694

Liesegang, T., Hoskins, H. Jr, Albert, D., O'Day, D., Spivey, B., Sadun, A., Parke, D. and Mondino, B. (2003) 'Ophthalmic education: where have we come from, and where are we going?' *American Journal of Ophthalmology* 136(1), pp. 114–121

Lowe, R. F. (1995) 'A history of primary angle closure glaucoma'. *Survey of Ophthalmology*, 40(2), pp. 163–170

MacKenzie, W. (1833) *A Practical Treatise on the Diseases of the Eye*. Boston, MA: Carter, Hendee and Co.

McKinnon, S. J. (2009) 'Glaucoma: current trends in diagnosis and treatment'. *Journal of Vision* 9(14), p. 16

Michelson, G. and Groh, M. J. (2002) 'Screening models for glaucoma'. *Current Opinions on Ophthalmology*, 12, pp. 105–111

Migdal, C. and Hitchings, R. (1991) 'The role of early surgery for open angle glaucoma' in J. Caprioli (Ed.) *Contemporary Issues in Glaucoma*. Ophthalmology Clinicians North America 4, pp. 853–9

Nasser, Q. J. and Stewart, W. C. (2006) 'Glaucoma treatment and diagnostic trends'. *Review of Ophthalmology*, 13(3), p. 87

Nelson, R. R., Buterbaugh, K., Perl, M. and Gelijns, A. (2011) 'How medical know-how progresses'. *Research Policy*, 40(10), pp. 1339–1344

Newell, F. W. (1997) 'The American Journal of Ophthalmology 1862–1864'. *Documenta ophthalmologica*, 93(1–2), pp. 135–143

Paton, D. and Craig, J. A. (1976) 'Glaucomas. Diagnosis and management'. *Clin Symp* 28 (2), pp. 1–47

Quigley, H. A. (1998) 'Current and future approaches to glaucoma screening'. *Journal of Glaucoma*, 7, pp. 210–219

Rachmiel, R., Trope, G. E., Chipman, M. L., Gouws, P. and Buys, Y. M. (2006) 'Effect of medical therapy on glaucoma filtration surgery rates in Ontario'. *Archives of Ophthalmology*, 124(10), p. 1472

Rafuse, P. (2003) *Adrenergic Antagonists. Glaucoma: Science and Practice*. New York: Thieme

Ramlogan, R. and Consoli, D. (2013) 'Dynamics of collaborative research medicine: the case of glaucoma'. *Journal of Technology Transfer*, (forthcoming)

Razeghinejad, M. R. and Spaeth, G. L. (2011) 'A history of the surgical management of glaucoma'. *Optometry & Vision Science*, 88(1), pp. E39-E47

Realini, T. (2011) 'A history of glaucoma pharmacology'. *Optometry & Vision Science*, 88(1), pp. 36–38

Ricci, B., Ricci, V. and Ziccardi, L. (2008) 'Surgical treatment for glaucoma in Italy: A five-year study period'. *The Internet Journal of Ophthalmology and Visual Science*, 6(1)

Rosenberg, N. (1976) *Perspectives on Technology*. Cambridge, UK: CUP Archive

Rosenberg, N. (2009) 'Some critical episodes in the progress of medical innovation: An Anglo-American perspective'. *Research Policy*, 38(2), pp. 234–242

Rossi, G. C., Pasinetti, G. M., Scudeller, L., Radaelli, R. and Bianchi, P. E. (2011) 'Do adherence rates and glaucomatous visual field progression correlate?' *European Journal of Ophthalmology*, 21(4), pp. 410–414

Schuman, J. S. and Lemij, H. G. (2000) *The Shape of Glaucoma: Quantitative Neural Imaging Techniques*. The Hague: Kugler Publications

Schwartz, G. F. and Quigley, H. A. (2008) 'Adherence and persistence in glaucoma therapy'. *Survey of Ophthalmology*, 53(S1), pp. 57–68

Sellem, E. (2000) 'Chronic glaucoma. Physiopathology, diagnosis, prognosis, principles of treatment'. *Rev Prat.*, 50, pp. 1121–5

Sharp, P.F., Manivannan, A., Xu, H. and Forrester, J. V. (2004) 'The scanning laser ophthalmoscope. A review of its role in bioscience and medicine'. *Physics in Medicine and Biology*, 49, pp. 1085–1102

Simon, H. A. (1969) *The Sciences of the Artificial*. Cambridge, MA: The MIT Press

Sommer, A., Quigley, H. A., Robin, A. L., Miller, N. R., Katz, J. and Arkell, S. (1984) 'Evaluation of nerve fiber layer assessment'. *Archives of Ophthalmology*, 102(12), pp. 1766–1771

Sommer, A., Tielsch, J. M., Katz, J. *et al.* (1991) 'Relationship between intraocular pressure and primary open angle glaucoma among white and black Americans. The Baltimore Eye Survey'. *Archives of Ophthalmology*, 109, pp. 1090–5

Stamper, R. L. (2011) 'A history of intraocular pressure and its measurement'. *Optometry & Vision Science*, 88(1), pp. E16-E28

Stone, E. M., Fingert, J. H., Alward, W. L. M., Nguyen, T. D., Polansky, J. R., Sunden, S. L. F., Nishimura, D., Clark, A. F., Nystuen, A., Nichols, B. E., Mackey, D. A., Ritch, R., Kalenak, J. W., Craven, E. R. and Sheffield, V. C. (1997) 'Identification of a gene that causes primary open angle glaucoma'. *Science*, 275, pp. 668–670

Traverso, C. E., Walt, J. G., Kelly, S. P., Hommer, H. A. and Bron, A. M. (2005) 'Direct costs of glaucoma and severity of the disease: a multinational long term study of resource utilisation in Europe'. *British Journal of Ophthalmology*, 89, pp. 1245–1249

Trick, G. L., Calotti, F. Y. and Skarf, B. (2006) 'Advances in imaging of the optic disc and retinal nerve fiber layer'. *Journal of Neuro-Ophthalmology*, 26(4), pp. 284–295

Trope, G. E. and Britton, R. (1987) 'A comparison of Goldmann and Humphrey automated perimetry in patients with glaucoma'. *British Journal of Ophthalmology*, 71(7), pp. 489–493

Tsai, J. C. (2009) 'A comprehensive perspective on patient adherence to topical glaucoma therapy'. *Ophthalmology*, 116(11), pp. 30–36 van der Valk, R., Schouten, J. S., Webers, C. A., Beckers, H. J., van Amelsvoort, L. G. and Schouten, H. J. (2005) 'The impact of a nationwide introduction of new drugs and a treatment protocol for glaucoma on the number of glaucoma surgeries'. *Journal of Glaucoma*, 14(3), pp. 239–242

Walland, M. J. (2004) 'Glaucoma treatment in Australia: changing patterns of therapy 1994–2003'. *Clinical Experimental Ophthalmol.*, 32(6), pp. 590–59

Weinreb, R. N. (2003) 'Weighing the cost of glaucoma progression'. *Ophthalmology Management*, August, pp. 191–204

Weller, P. (1826) *Theoretical and Practical Treatise of Eye Disease*. London: Grattan

Wiggs, J. L. (2007) 'Genetic etiologies of glaucoma'. *Archives of Ophthalmology*, 125(1), pp. 30–37

Wilson, M. R. and Gaasterland, D. (1996) 'Translating research into practice: controlled clinical trials and their influence on glaucoma management'. *Journal of Glaucoma*, 121, pp. 139–46

Wise, J. B. and Witter, S. L. (1979) 'Argon laser therapy for open angle glaucoma'. *Archives of Ophthalmology*, 97, pp. 319–22

Zimmerman, T. J. (1977) 'Timolol maleate - a new glaucoma medication?' *Investigative Ophthalmology & Visual Science*, 16(8), pp. 687–688

Zimmerman, T. J., Kooner, K. S., Kandarakis, A. S. and Ziegler, L. P. (1984) 'Improving the therapeutic index of topically applied ocular drugs'. *Archives of Ophthalmology*, 102, pp. 551–553

Conclusion

The case studies contained in this volume offer a broad assortment of perspectives on medical innovation and bring to the fore a number of common characteristics of the process under analysis. First, the development of a new medical practice usually does not occur over short time spans but, rather, in most cases over relatively long periods. Both the material contained here as well as other case studies on the subject matter show that this is because the know-how that is relevant to the effect of restoring human health requires the vision, experience and expertise of different professionals, and of the institutions in which they operate. Accordingly, the learning process that drives the accumulation and cyclical adaptation of relevant know-how entails considerable efforts in the orchestration of different capabilities, perceptions and interests. A second prominent feature that stands out from reading the chapters is that the process leading to successful medical innovation involves significant uncertainty. This may regard not only the way in which a clinical problem ought to be addressed but also the very identification of what the problem is, and whether the particular remedies that are being developed are viable, effective and sustainable. As a matter of fact, tackling health issues entails working around the boundaries of that penumbra of uncertainty. It is therefore not surprising that the assessment of the reliability of a new clinical procedure is often compared with the trial and error learning that is typical of disciplines like engineering where experience is a key ingredient in the definition of workable solutions. Third, at times the achievement of effective responses to a medical problem occurs in the absence of a complete fundamental understanding of the disease. The contrary argument, that stronger scientific understanding invariably guides progress in medicine along a paradigm – borrowing from Kuhn's terminology – leads to a mischaracterisation. Quite often improvements in practice stem from ways of understanding and from research routines that are developed with rather narrow goals in mind. If anything, the development of new diagnostic and therapeutic modalities are better understood as being own-standing paradigms, at times unfolding along competing directions. This is not to say that searches are random: they are 'guided' searches although their outcome is very rarely predictable *ex ante*. Also, this is not to lessen the importance of the path-breaking discoveries that originate from outside the realm of medicine and that have enabled better practice on several occasions. The last crucial feature of medical innovation that emerges from these chapters is that, while the conception and design of more effective treatments occurs at some distance from actual practice, there is a considerable amount of learning by doing and by using involved. This at times entails that the observation of successful outcomes acts as a guide for new biomedical research on the underlying causes of disease. But, regardless of whether scientific understanding is strong or weak, all the case studies in this collection provide ample evidence of strong feedback between adoption and research and development in medicine, whereby new technologies need to be continuously adapted because drawbacks (for example, side-effects) and potential improvements become apparent only through use. Crucially, the way in which both artefacts and practices evolve depends on the exchange of knowledge between clinicians, hospital administrators, patients, insurers and regulators, who influence the rate and direction of medical innovation by explicitly identifying priority needs and redefining modes of service provision.

In consideration of the features just outlined, the case studies adopt a dynamic perspective on medical innovation based on the notion that scientific understanding, technology and clinical practice co-evolve along the coordinated search for solutions to medical problems. The different chapters follow an historical approach to emphasise that the advancement of medical know-how is a contested, nuanced process, and that it involves a variety of knowledge bases whose evolutionary paths are rooted in the contexts in which they emerge. One prominent advantage of this particular analytical lens is that the nexus between the technological and institutional dimensions of innovation in healthcare has been brought clearly to the fore. Following Rosenberg's intuition (1994), this allows us to treat the development and use of innovation as twin processes that unfold over time, mutually shaping each other as learning by using expands or narrows the technology's scope of application. These overarching characteristics emerge forcefully in the narrative of the cases that make up this volume.

The conceptualisation of medical innovation as the exploration of design space is clearly illustrated by the case of the intra-ocular lens (IOL). This design space is defined by the perception of clinical problems and constrained by the technological competences that can be mobilised at specific points in time over the long-term development of new solutions to these problems. The IOL innovation-diffusion process took 40 years to unfold from the first implant to the establishment of a standardised procedure now diffused on a large international scale. Here too scientific understanding and clinical procedures are progressively refined through experience. Problem-solving stimulated improvements in practices and extended their scope of application in ways that challenged previous understanding. This generated progress along trajectories of change whose driving force was the path-dependent growth of knowledge. This knowledge not only grew but also fundamentally changed over time through incremental and complementary steps, and this transformation was associated with changes in the social networks in which innovation is embedded. The IOL case also draws attention to the institutionalisation of demand as an intrinsic aspect of healthcare innovation. In the innovation-diffusion process the level and attributes of localised demand is key to the shaping of the division of innovative labour, and the relation between demand and medical need is itself dependent on the prevailing clinical technology and health care management practices.

The development of coronary angioplasty illustrates the strong complementarities between clinical understanding and practical skills, between radical breakthroughs and the incremental innovations that made the new procedure clinically effective and economically efficient, and the interconnected role of advances in know-how and the institutional context of innovation. This particular case also highlights how, as science, technology and organisations co-evolve, the nature of the problems being addressed gradually changes. As solutions to some problems are found, others come into view, which are often generated by the very solutions introduced earlier on in the innovation or diffusion process. Moreover, as new problems present themselves, opportunities for further change emerge in the type and composition of relevant technological knowledge and its market applications. In the domain of interventional cardiology medical progress has followed specific trajectories of change, which unfolded over time in fundamentally uncertain ways and converged and diverged along sequences of problems and solutions. These are observable *ex post* as the result of an evolutionary process of variation and selective retention of know-how that expert knowledge and clinical use identified as effective.

The development of left ventricular assist devices (LVADs) provides an interesting counterpoint to the development of coronary angioplasty. The use of LVAD for the treatment of advanced heart failure is not as well established as the use of PCI for coronary artery disease, although the two technologies and associated procedures developed over comparable periods of time. Experimentation with new design, components, and materials generated remarkable improvements over the early models and led to a *de facto* standard for first-generation LVADs. In the early 1990s commercial incentives for further experimentation grew once proof of concept was established for the new technology and with the support of federal funding through the NIH's SBIR programme. Improvements in use, complementary procedures and market competition led to a proliferation of designs whose comparative performance is still being evaluated against persistent barriers to the diffusion of this technology, not least its costs. Not unlike the case of coronary devices, many improvements in the new technology cannot be attributed to fundamental advances in understanding of heart disease but, instead, to trial-and-error learning, facilitated in both cases by the introduction of a dedicated clinical data registry.

Interestingly, the experiences generated in the context of using the artifacts illustrated so far have elucidated important aspects of the attendant medical problem. This is not so in the case of the Bryan cervical artificial disc, a device designed for relieving the symptoms of degenerative disc disease. As an innovation it offers a serious challenge to the preeminence of the gold standard disc fusion technology in treatment of this disease. Although interest in artificial discs existed since the 1960s, early failures dampened enthusiasm for the idea as treatment trajectories turned to disc fusion. However, by the 1980s artificial discs were back on the agenda and designed around two different trajectories, one based on a ball-and-socket articulation between rigid materials and the other concerned with replicating the articulation properties of the disc and its load absorption properties. While the latter would intuitively seem to be a more sensible design, the importance of load absorption is highly contested in both design trajectories, particularly because of the absence of empirical data. Such uncertainty is compounded by the lack of studies addressing the differential performance of these competing designs. Nevertheless, the case study based on the Bryan disc represents a third trajectory in the design space, one that incorporates both competing operational principles in one artifact. Interestingly, the Bryan disc has enjoyed considerable success in a relatively short space of time. Such technological hybridisation results from problem-solving strategies constrained by uncertain knowledge spaces and can be viewed from a context of incremental innovation in which emergent pathways co-exist with, rather than abruptly closing off, old ones. If the goal of this technology is to preserve and restore the natural movement to the cervical spine, and if the clinical outcomes from such devices exceed those from fusion technology, then at the very least a new treatment standard will have emerged.

The case study on diagnostic innovation of cervical cancer illustrates the role of institutional factors in guiding the interplay between competing technologies in the US. The incumbent cervical cytology or pap smear testing has been a cornerstone of preventive health policy since the 1940s while the challenger, a new molecular diagnostic test for the human papilloma virus (HPV test), has already moved from a niche player as a triage test for sometimes ambiguous pap smear results in the 1980s, to a more central role as an adjunctive test along with the pap smear, and with ambitions to becoming the primary screening test. The contextualisation of the development of these two technologies sharply diverges. While non-profit organisations and academic researchers largely carried out pap testing, HPV testing typifies a relatively new trend in which diagnostic companies are becoming increasingly important in the development and diffusion of innovative molecular diagnostics. As a relatively small and young diagnostic company, Digene (eventually acquired by Qiagen in 2007) pursued a commercial strategy that included patenting to gain market exclusivity, marketing direct to physicians, direct to consumer advertising and investing heavily in research creating an international network through which the clinical utility of its Hybrid Capture (HC) test was demonstrated and championed. Thus, in leading the commercial field, Digene's HC test quickly became tightly associated with the emerging technology of HPV testing.

Weakness of scientific understanding again plays heavily in the early discourse on the discovery of the polio vaccine. This development shares with other case studies in this volume the unpredictable twists and turns characteristic of the evolution of knowledge. Vaccine development is framed as a learning process where the accumulation of technological knowledge takes centre stage. While the proof of concept for the vaccine came as early as 1910, optimism for its development quickly waned and it was not until 40 years later that the Salk vaccine (and later the Sabin vaccine), would become available. In the interim, various blind alleys were encountered as researchers struggled to understand the pathology and epidemiology of the disease as well as how a vaccine would work. Early attempts to implement a vaccine based on an existing model tragically resulted in the deaths or paralysis of thousands of children and this failure refocused scientific and public policy efforts. The outcome of this was the strengthening of testing regimes, the development of new methods of tissue culturing and subsequently confirmation of the existence of different strains of poliomyelitis, a finding that would provide the standard against which candidate vaccines could be compared. Overlooking these developments was the establishment of a governance structure in the form of the National Foundation for Infantile Paralysis that provided the coordination of the research, testing activities and ensured that the knowledge growth was cumulative and shared.

The last chapter on glaucoma reemphasises the uncertainty that epitomises the discovery process in many areas of medical research and practice. It is a classic case in which the unfolding of knowledge about the causal factors involved in a complex disease (or set of diseases to be more precise) has been slow and this has led, for a considerable time, to an inadequate framing of the disease and consequently the therapeutic options possible. While there has been a number of advancements in diagnostic technologies that contributed to a growing understanding of the disease, so far, despite the best efforts of the scientific community in glaucoma, there remains a gap between 'what is known about the disease and what can be done about it'. Treatment regimes are able to slow down the progression of the disease but they are unable to reverse progressive damage to the visual field. In contrast to the other cases in this volume, glaucoma remains a disease very much in search of a 'cure'. The discovery of a correlation between the erosion of the optic nerve and visual field loss in the 1970s shifted the thinking about glaucoma from the then prevailing hydraulic blockage model in favour of glaucoma as a neuropathy and paved the way for the interaction of glaucoma with new fields of research such as neurology, cellular biology and genetics, interactions that lend hope for a viable solution. While these new disciplinary combinations brought about new understandings about the biological processes that are involved in glaucoma, there remains a formidable gap between knowledge of the disease and viable therapies. To reprise a key message, this case is a 'forceful reminder of the extent to which "knowledge permits" but "also constrains"'.

To conclude, the history of medicine seen through the lenses of the case studies of this volume emerges as rich in examples of ways in which the path that leads to improving the 'way of doing things' emerge. While charismatic pioneers, be they 'hero-surgeons', visionary scientists or disruptive entrepreneurs, can provide formidable creative impetus, it is systemic change that transforms the paths to successful treatments on a large scale. The cases illustrate that organising a system for the provision of healthcare services spans a remarkably wide number of activities and, thus, of skills and forms of technical and practical knowledge.

This has at least two important implications. Firstly, this complex system requires effective mechanisms of coordination to achieve the overall goal of patient care, a point that reinforces our emphasis on the institutional features of innovation. Secondly, the search for and the implementation of new clinical solutions guide the redistribution of knowledge across areas of specialisation and, more cogently, the opportunities for further connections within and across domains of knowledge and practice, which shape the division of innovative labour over time. As the cases have shown, some aspects of medical innovation are specific to the clinical problems at hand; others are more general and provide more general lessons about the way in which medical knowledge evolves. The advancement of medical know-how and its clinical application is a process characterised by uncertainty and unevenness in determinants and outcomes. It is a process of punctuated change, with recurrent long periods of slow progress and moments of rapid advancement after unpredictable breakthroughs, realised through both new knowledge creation and successful refinement or recombination of existing knowledge.

Index